Algal symbiosis

Few groups of organisms have been more successful in forming intimate symbioses with other organisms than the evolutionarily diverse group, the algae. Within every division of these organisms, and in every community they inhabit, symbiotic interactions have evolved, in some cases having profound effects on the ecosystem.

Algal symbioses form a continuum, each interaction being a function of the evolutionary history of the separate "players" as well as of the partnership. Each algal symbiosis is unique; although symbioses may be described, they cannot necessarily be categorized. In fact, it is arguable that attempts to do so may actually obscure the true physiological and genetic nature of the interaction, and quite possibly bias the scientific objectivity necessary for the required experimental quantitative and qualitative studies of the association.

Symbiotic systems provide biologists with extremely useful experimental tools to study important biological phenomena. No longer should research in this area be considered "exotic"; rather, it is central to the understanding of cell biology and the origins of innovation in evolution.

Algal symbiosis

A continuum of interaction strategies

Edited by
LYNDA J. GOFF
Department of Biology and the
Center for Coastal Marine Studies
University of California, Santa Cruz

CAMBRIDGE UNIVERSITY PRESS
Cambridge
London New York New Rochelle
Melbourne Sydney

CAMBRIDGE UNIVERSITY PRESS
Cambridge, New York, Melbourne, Madrid, Cape Town,
Singapore, São Paulo, Delhi, Tokyo, Mexico City

Cambridge University Press
The Edinburgh Building, Cambridge CB2 8RU, UK

Published in the United States of America by Cambridge University Press, New York

www.cambridge.org
Information on this title: www.cambridge.org/9780521177429

First published 1983
First paperback edition 2011

A catalogue record for this publication is available from the British Library

Library of Congress Cataloguing in Publication data
Main entry under title:
Algal symbiosis
Includes bibliographies and indexes.
1. Algae – Ecology – Congresses. 2. Symbiosis – Congresses. I. Goff, Lynda J.
QK565.A38 1983 589.3′52482 83–7275

ISBN 978-0-521-25541-7 Hardback
ISBN 978-0-521-17742-9 Paperback

Contents

Acknowledgments

This volume is an outgrowth of a symposium presented at the third combined meeting of North American Botanists (Botany 80) in Vancouver, British Columbia, during July 1980. The meeting was sponsored jointly by the Canadian Botanical Association, the Botanical Society of America, the Phycological Society of America, the American Bryological and Lichenological Society, and the American Fern Society. The Phycological Society of America was the direct sponsor of the symposium on algal symbiosis.

The contributors to this publication were selected to provide a broad coverage of the continuum of interactions apparent in algal symbiosis. The text is not intended to be a comprehensive coverage of algal symbiosis, but rather a major sampling of the interactions on which experimental work has been undertaken. Other topics that would have been logically included, such as the cyanellae symbiosis, the biology of parasitic red algae, zoochlorellae, and protozoans, have been reviewed recently, and the reader is directed to such texts as *Progress in Phycological Research* (Round, F. E., and Chapman, D. J., eds., 1982), *Cellular Interactions in Symbiosis and Parasitism* (Cook, C. B., Pappas, P. E., and Rudolph, E. D., eds., 1980), *Symbiosis in Cell Evolution* (Margulis, L., 1981), and the article by M. Richmond and D. C. Smith entitled *The Cell as a Habitat* (1979, Proceedings of the Royal Society of London), as well as those in the 1975 *Symposium Proceedings on Symbiosis* published by the Society of Experimental Biology.

I extend my sincere thanks to the contributors for their excellent articles. Each manuscript was reviewed by two or three individuals and their critical comments and input were of great help to the authors and editor. Thanks are also extended to Dr. Robert DeVreede (University of British Columbia), who provided assistance during the symposium, and to Dr. Janet Stein (University of British Columbia), for her valuable editorial advice. The editorial staff of Cambridge University Press has been extremely helpful, and the assistance of Dr. Richard Ziemacki, Ms. Katie Gontrum, and Ms. Mary Byers was considerable. The research and indexing assistance of Mr. David Lerner and Ms. Eugenia McNaughton (University of California, Santa Cruz) was invaluable, as was the clerical and financial support provided by the Department of Biology and the Center for Coastal Marine Studies of the University of California, Santa Cruz.

Last, I must thank the many bright and eager students who have contributed greatly to classes that I have taught on algae and algal symbiosis at the University of California. Their enthusiasm and often unique perspectives continue to open up new ways of looking at very old symbioses!

Lynda J. Goff

University of California, Santa Cruz

Introduction

LYNDA J. GOFF

Department of Biology and the Center for Coastal Marine Studies
University of California, Santa Cruz
Santa Cruz, CA 95064

> Nature abhors a pure culture; symbioses, whether ecto- or endocellular, are the rule rather than the exception.
>
> Lynn Margulis, 1980

> If it is the nature of living things to pool resources, to fuse when possible, we would have a new way of accounting for the progressive enrichment and complexity of form in living things.
>
> Lewis Thomas, 1974

Biologists' perceptions of the interactions of organisms have changed profoundly during the past century. No longer is nature viewed as "red in tooth and claw," where only the strongest and most aggressive survive and dominate. Rather, studies of the complexity of organismal interactions within communities have clearly revealed that success is often linked to the ability of an individual, or a population of individuals, to interact intimately with other organisms within the community.

Selective advantage is attained through exploiting other members of the community. However, the exploitation is generally far more subtle than merely eating one's competitor. The fungi of lichen associations attain a selective advantage by capturing and enslaving the photosynthetic capabilities of their algal partners. In these highly refined interactions, the association is one of "controlled parasitism" (Smith 1980), in which the fungus exploits the alga but has little deleterious effect on the alga. Mutual exploitation between two members of a community may also evolve, to the advantage of both individuals. This is the case in some associations of invertebrates and algae, and in mycorrhizal associations of fungi and higher plants.

DeBary (1879) employed the term *symbiosis* (*sym*, together; *biosis*, living) to describe the intimate interactions of dissimilarly named organisms ("Zusammenleben ungleichnamigen Organismen") and, contrary to subsequent usage (Lewin 1982), indicated that symbiosis includes all degrees of parasitism as well as mutualistic and commensalistic associations (Goff

1982). A more current definition of symbiosis is an "association, for significant portions of the life cycles, of individuals that are members of different species" (Margulis 1981).

Since deBary introduced the term *symbiosis*, literally thousands of symbiotic associations involving plants, animals, fungi, protistans, and prokaryotes have been described. Probably few groups of organisms have been more successful in forming intimate symbioses with other organisms than the evolutionarily diverse group of organisms known as the algae. Within every division of these organisms, and in every community they inhabit, symbiotic interactions have evolved, and in some cases these have had profound effects on the ecosystem. E. J. Ferguson-Wood (1967) states,

> In the tropics, many, probably most of the recurring animals have algal symbionts and these extend through most of the invertebrate phyla . . . Quantitatively it is probably true that the symbiotic algae are more important than the phytoplankton and the free benthic algae in coral reefs and other shallow waters with calcareous sediments, and the productivity of such waters is essentially due to such symbionts, which are more usually known as zooxanthellae.

Algae occur as symbiotic extracellular (intercellular, exhabitational) or intracellular (inhabitational) associates of vertebrate, invertebrate, protistan, fungal, angiosperm, and algal hosts. In the extreme case, only the photosynthetic organelle of the alga, the plastid, is incorporated into the host's cytoplasm, where it remains active photosynthetically, providing fixed carbon to the heterotrophic host.

In many symbiotic associations, the algal symbiont's morphology is altered, as is its biochemical and physiological activity. As the symbiosis becomes progressively more obligate, a loss of capabilities for an independent life becomes evident. This may involve the loss of genetic information necessary for autonomy. In the case of the symbiosis between the cryptomonad-like flagellate *Cyanophora paradoxa* and its cyanobacterial endosymbiont *Cyanocyta korschikoffiana*, pigment synthesis (c-phycocyanin) in the cyanelle inclusion is under partial control of the host's ribosomes (Siebens and Trench 1978, Trench and Siebens 1978). This situation may be analogous to the synthesis of ribulose bisphosphate carboxylase in plant cells, where both the nuclear and plastic genomes are necessary for enzyme synthesis.

During the past 25 years, nearly 1,500 papers have been published dealing with algal symbiosis. However, many of these papers are primarily descriptive and, as stated by Margulis (1981), "are often treated in the biological literature as exotic." Until recently, experimental studies of algal symbioses have been few and for the most part have been undertaken to classify the interaction – that is, to determine if the association is parasitic, mutualistic, commensalistic, biotrophic, necrotrophic, exhabitational, inhabitation, obligate, facultative, specific, permissive, transient, or persistent (Henry 1966, Lewis 1973, Starr 1975, Lewin 1982). However, biological phenomena involving organismic interactions are rarely abruptly discontinuous (Scott 1969, Smith 1980, Starr 1975, Goff 1982). More often they exist in a more-or-less smooth and continuous gradation or continuum.

As will be evident in the following chapters, algal symbioses are often difficult or impossible to classify. They form a continuum, each interaction being a function of the evolutionary history of the separate "players" as well as the partnership. This then is the basic theme of this volume. Each algal symbiosis is unique; although symbioses may be described, they cannot necessarily be categorized. In fact, it may be argued that attempts to do so may actually obscure the true physiological and genetic nature of the interaction, and quite possibly bias the scientific objectivity necessary for the required experimental quantitative and qualitative studies of the association.

The second theme of this book is that symbiotic systems provide biologists with extremely useful experimental tools to study important biological phenomena. For example, research on the lichen symbiosis (Ahmadjian and Jacobs, Chapter 8) is providing insight into intercellular transport mechanisms (apoplastic and symplastic), whereas intracellular (interorganelle) transport mechanisms are being examined in endosymbiotic associations of zooxanthellae and zoochlorellae and their invertebrate hosts (Pardy, Chapter 1; Taylor, Chapter 2; Lee and McEnery, Chapter 3; Anderson, Chapter 4), and in the association of the mollusc *Elysia* and its endosymbiotic algal chloroplasts (Hinde, Chapter 6).

In addition, algal symbiotic associations provide the means to examine (1) how different genomes may function in conjunction and how such an interaction is integrated, (2) the basis of cell – cell recognition mechanisms (Chapman and Good, Chapter 9; Ahmadjian and Jacobs, Chapter 8; Peters and Calvert, Chapter 7; Hinde, Chapter 6; Lee and McEnery, Chapter 3; Taylor, Chapter 2; Anderson, Chapter 4), (3) mechanisms by which new species, and through them higher taxa, originate, and (4) how the eukaryotic cell may have evolved. No longer should research in this area be considered "exotic"; rather, as stated by Margulis (1980), it is central to the understanding of cell biology and the origins of innovation in evolution.

References

deBary, A. (1879). *Die Erscheinungder Symbiose.* Verlag von Karl J. Trubner, Strassburg.

Ferguson-Wood, E. J. 1967. *Microbiology of Oceans and Estuaries.* Elsevier Oceanography Series, vol. 3. Elsevier, Amsterdam.

Goff, L. J. (1982). Symbiosis and parasitism: Another viewpoint. *BioScience,* **32,** 255–6.

Henry, S. M. (1966). Association of microorganisms, plants and marine organisms. In: *Symbiosis,* vol. 1, ed. S. M. Henry. Academic Press, New York.

Lewin, R. A. (1982). Symbiosis and parasitism – definitions and evaluations. *BioScience,* **32,** 254–9.

Lewis, D. H. (1973). Concepts of fungal nutrition and the origin of biotrophy. *Biological Reviews* **48,** 261–78.

Margulis, L. (1980). *Symbiosis as parasexuality.* In: *Cellular Interactions in Symbiosis and Parasitism,* ed. C. B. Cook, P. W. Pappas, and E. D. Rudolph, pp. 263–73. Ohio State University Press, Columbus.

– (1981). *Symbiosis in Cell Evolution.* W. H. Freeman & Co., San Francisco.

Scott, G. D. (1969). *Plant Symbiosis.* Edward Arnold, London.

Siebens, H. B., and R. K. Trench (1978). Aspects of the relation between *Cyanophora paradoxa* (Korschikoff) and its endosymbiotic cyanelles *Cyanocyta korschikoffiana* (Hall and Claus). III. Characterization of ribosomal ribonucleic acids. *Proc. R. Soc. Lond.* (B) **202,** 463–72.

Smith, D. C. (1980). Mechanisms of nutrient movement between the lichen symbionts. In: *Cellular Interactions in Symbiosis and Parasitism,* ed. C. B. Cook, P. W. Pappas, and E. D. Rudolph, pp. 197–227. Ohio State University Press, Columbus.

Starr, M. P. (1975). A generalized scheme for classifying organismic associations. *Symp. Soc. Exp. Biol.* **29,** 1–20.

Trench, R. K., & Siebens, H. B. (1978). Aspects of the relation between *Cyanophora paradoxa* (Korschikoff) and its endosymbiotic cyanelles *Cyanocyta korschikoffiana* (Hall and Claus). IV. The effects of rifampicin, chloramphenicol and cycloheximide on the synthesis of ribosomal ribonucleic acids and chlorophyll. *Proc. R. Soc. Lond.* (B) **202,** 473–83.

1

Phycozoans, phycozoology, phycozoologists?

R. L. PARDY

School of Life Sciences
University of Nebraska
Lincoln, Nebraska 68588

Experimental and quantitative studies on several algal-invertebrate symbioses have suggested or shown directly that these associations possess a variety of interacting systems that stabilize and perpetuate them. It is becoming evident that host-algal symbiont interactions occur at many levels of biological organization. Among some contemporary workers in algal-invertebrate symbioses there has been a shift away from the "reciprocal benefit," the "who does what for whom," and the catalog-categorization approaches (Starr 1975) to symbiosis research. There is a growing emphasis on studies probing the underlying mechanisms allowing or causing animals and algae to form stable entities that thrive and persist through time. We are becoming aware that algal-animal symbioses have unique biological identities. The association is an organism that makes its living in a particular way, in a particular ecological context.

Various aspects of algal-invertebrate symbiosis have been reviewed repeatedly in the last 10 years (Cook 1980, Muscatine 1974, Muscatine et al. 1975, Smith 1974, Smith et al. 1969, Trench 1979). My goal is not to present still another review but rather to offer a point of view and to suggest a paradigm that might help unify and enhance our understanding of algae-invertebrate symbioses. In the following discussion I use the words *symbiosis, host, symbiont*, and *aposymbiont*. These terms are used operationally, with symbiosis taken to mean the intimate physical association of unicellular algae and animals; with host and symbiont referring to the animal and algal components, respectively; and with aposymbiont referring to an invertebrate host disassociated from its algal symbionts. I also avoid the terms *mutualism, benefit, partners*, and other symbiosis jargon (except as previously noted).

The word *parasite* conjures up an image of an organism having a particular lifestyle. The word does not imply any particular systematic or phylogenetic grouping, but rather a functional category of organisms.

This paper was prepared while the author was supported by NSF Grant No. PCM-7904224.

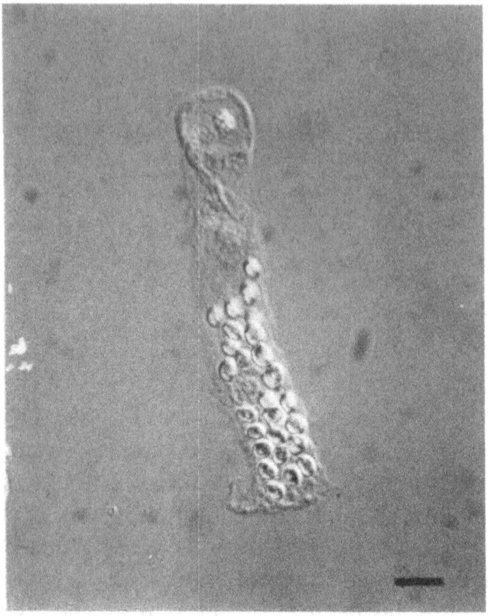

Fig. 1. Isolated digestive cell from *Hydra viridis* showing symbiotic algae (green spheres). Bar equals 10 μm, Nomarski optics.

Herbivore, xerophyte, annual, and *ectotherm* are other useful terms describing or grouping organisms of similar habits. I wish to introduce the term *phycozoan* (*phyco*, seaweed, *zoa*, animal) to denote the compound organism resulting from the intimate association of algae and animals. Phycozoans are symbioses where algal unicells (phycobionts = algal symbionts) live inside an animal (zoobiont = animal host), within its cells (Fig. 1), and/or among its tissues (Fig. 2). Because I intend phycozoa to be a functional category, I would consider the algal-bearing protists to be phycozoans. Occasionally algal plastids are symbionts as in the case of certain marine slugs that acquire and harbor chloroplasts from siphonaceous algae (see Trench 1975, for review and Chapter 6 in this volume). These are also phycozoan associations.

The phycozoan is an integrated amalgamation of (1) characteristics specific to and maintained by the associating bionts, (2) preadaptations that are modified, amplified, and recruited by the phycozoan, and (3) new features found only in the phycozoan or resulting from association. In these respects phycozoans are analogous to lichens (see Chapter 8 in this volume). The lichen thallus emerges as the biological unit having specialized characteristics and adaptations not found in free-living (not lichenized) fungi and algae. Fungi and algae making up a lichen thallus do not lose their identity even though the morphology and physiology of the thallus may be very complex. In phycozoans, the phycobionts invariably retain the ability to photosynthesize. Moreover, although free-

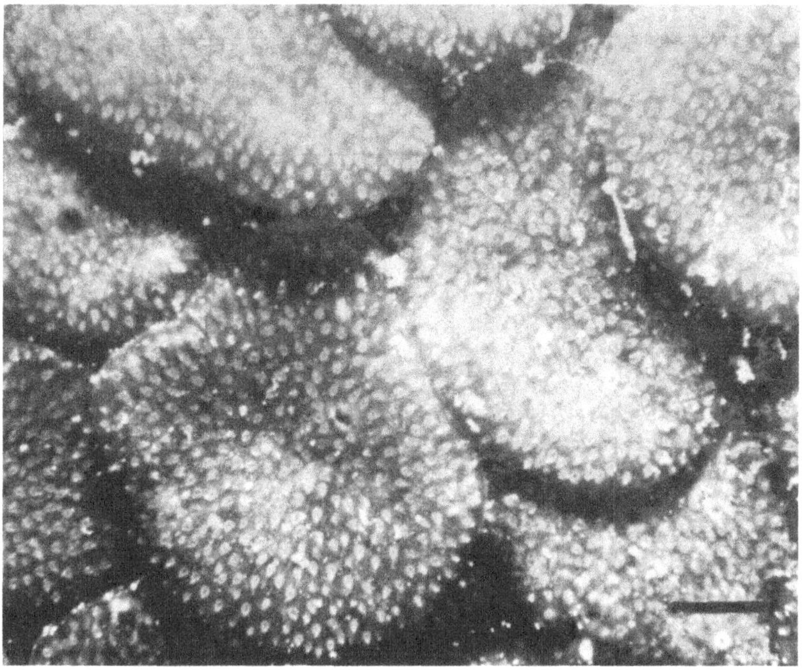

Fig. 2. Colonies of *Trididemnum*, a colonial ascidian containing *Prochloron* cells. Bar equals 1 cm. Photograph courtesy of Dr. Charles Berkland, University of Guam.

living algae are known to leak or excrete organic substances (Fogg 1962), in symbiotic algae leakiness (a preadaptation?) is amplified such that large amounts of specific organic molecules are released. Finally, after becoming part of a phycozoan, symbiotic algae may undergo a variety of modifications such as the loss of flagella, modification of cell walls (Oschman 1966), and development of a new ensemble of surface antigens (Pool 1979).

Phycozoans have emergent properties distinguishing them from simply animals with algae in them. These properties enable the phycozoan to exist and thrive in niches from which its component bionts, as individuals, are excluded. Hence phycozoans may define an ecological niche (Hutchinson 1959) though presently little information exists on this aspect of phycozoan biology.

It seems reasonable to recognize phycozoans as stable biological units, real organisms, analogous to lichens, upon which the forces of evolution act. Mutations in either phycobiont or zoobiont have the potential for increasing or decreasing the fitness of the phycozoan. A novel possibility, arising either from mutation or immediate environmental factors, is that the phycozoan becomes extinct, even while one or both of its component bionts persist, though unassociated.

Table 1. *Diversity of phycozoans*

Phylum	Examples: Phycobiont	Examples: Zoobiont
Protozoa	*Chlorella*	*Paramecium bursaria*
	Chlorella	*Stentor polymorphous, Ophrydium, Vorticella*
	Symbiodinium	Radiolarians Foraminiferans
Porifera	*Chlorella*	*Spongilla* sp.
Coelenterata	*Chlorella*	*Hydra viridis* (green hydra)
	Symbiodinium	*Anthopleura* sp. (sea anemone)
	Symbiodinium	*Pocillopora* sp. (coral)
	Symbiodinium	*Cassiopeia* sp. (jellyfish)
Platyhelminthes	*Chlorella*	*Dalyellia viridis* (freshwater flatworm)
	Platymonas	*Convoluta roscoffensis* (marine flatworm)
Mollusca	*Symbiodinium*	*Tridacna crocea* (giant clam)
	Vaucheria chloroplasts	*Elysia viridis* (slug)
	Chlorella	*Anodonta* (freshwater clam)
Chordata	*Prochloron*	*Diplosoma virens* (compound tunicate)

Although a fairly wide assortment of invertebrates form phycozoans, comparatively few species of algae become involved. Table 1 is not an exhaustive catalog, but it illustrates the diversity of animals and kinds of algae that form phycozoans. Despite the apparent variety and various combinations of algae and invertebrates, all phycozoans seem to display the following characteristics:

1. The transfer of photosynthetically produced organic substances from the phycobionts is essential for the stability and persistence of the phycozoan.
2. Because of (1), photosynthesis is a central and critical aspect of the phycozoan's biology.
3. Phycozoans have mechanisms that assure continuity of the association through successive generations of the zoobiont.

Phycobionts from all phycozoans so far analyzed translocate substantial quantities of soluble organic molecules to the zoobiont cells and also release products to the medium when analyzed in vitro (Table 2 gives some examples). Experiments with ^{14}C have shown that photosynthetically reduced carbon produced by phycobionts enters the zoobiont's metabolic network, eventually appearing in all major biochemical fractions (Muscatine & Cernichiara 1969, Pardy 1980, Trench 1971). Using

Table 2. *Translation by phycobionts from various symbionts*

Phycobiont	Zoobiont	% of photosynthesis translocated	Major product(s)	Reference
Platymonas	*Convoluta*	8	Ala, Gly, Pyr	Muscatine 1974
Chlorella	*Hydra*	40	Maltose	Cernichiara et al. 1969
Chlorella	*Paramecium*	15–86	Maltose	Muscatine et al. 1967
Chlorella	*Spongilla*	4	Glucose	Muscatine et al. 1967
Chlorella	Mutant *Hydra*		Glucose	Muscatine et al. 1967
Symbiodinium	*Rhizostoma*	20	Glycerol	Trench 1971
Symbiodinium	*Cassiopeia*	23	Glycerol	Trench 1971
Symbiodinium	*Anthopleura*	49	Glycerol	Trench 1971
Symbiodinium	*Aiptasia*	35	Glycerol	Trench 1971
Symbiodinium	Fungia	25	Glycerol	Trench 1971
Symbiodinium	*Zoanthus*	42	Glycerol	Trench 1971
Symbiodinium	*Tridacna*	40	Glycerol	Muscatine 1967
Symbiodinium	*Pocillopora*	40	Glycerol	Muscatine 1967

green hydra and indirect calorimetry, Pardy and White (1977) showed that the flow of carbohydrate from the phycobiont was sufficient to induce a pronounced carbohydrate metabolism in the phycozoan. By comparison, aposymbiotic hydra exhibit a fat metabolism almost exclusively. Similar findings are reported for phycozoan sea anemones. Apparently this algal-to-animal flux of nutrients is the fundamental basis for the existence of phycozoans. It allows the zoobiont to inhabit environments that experience continuous or periodic shortages of exogenous food or to invade habitats usually foreclosed to them for lack of normal food items. In reef-forming corals mounting evidence points to the translocation of nutrients from phycobiont to zoobiont as being the major factor in promoting rates of calcification sufficiently high for reef building and maintenance (Muscatine & Cernichiari 1969, Muscatine & Porter 1977). If photosynthesis is inhibited, the zoobiont often experiences reduced growth rates and/or degenerates. Although aposymbiotic zoobionts of some phycozoans can be artifically produced and maintained in the laboratory (*Hydra*: Whitney 1907, Muscatine & Lenhoff 1965, Pardy 1976; *Paramecium bursaria*: Karakashian 1963; *Convoluta roscoffensis*: Provasoli et al. 1968, Boyle & Smith, 1975), such forms are rarely found in nature, probably because they cannot survive. The point is that the existence and well-being of the phycozoan depends upon a continuous or intermittent supply of nutrients provided by the phycobionts. Because the organic nutrients released by the phycobionts originate from photosynthesis, photosynthesis is a critical process upon which the stability of the phycozoan depends.

To photosynthesize at a useful rate, phycobionts require optimum supplies of carbon dioxide and photons. This means the phycobionts must be strategically located within the phycozoan so that carbon dioxide flow and light penetration are maximized. Hence, design limitations and requirements are placed on phycozoan structure. In most phycozoans the phycobionts are located in or among tissues near the host's surface. In coelenterate phycozoans the phycobionts occupy the gastrodermis, a tissue seldom more than a few millimeters from the animal's external surface. In worm and slug phycozoans, the phycobionts are often associated with the gut, which has many diverticulae or tubules ramifying throughout the animal and lying close to the surface (Dorey 1965, Oschman 1966). A remarkable adaptation occurs in the mantle of the giant clam, *Tridacna*. The phycobionts occupy cavities in the mantle tissue, which is extruded between the valves and comes partially to overlay the external surface of the clam. Imbedded in the mantle are lenslike structures that facilitate the penetration of light to the phycobionts. Finally, in a fascinating analysis, Porter (1976) has shown a relationship between the morphology of phycozoan reef corals and the degree of autotrophy they exhibit. Those corals with the highest degree of autotrophy have surface areas maximized for efficient photon capture.

In addition to structural aspects, zoobionts have certain behaviors that enhance or modulate phycobiont photosynthesis. In the anemone *Aiptasia*, the density, hence the degree of mutual shading, of phycobionts can apparently be regulated by the degree of tentacle contraction (Glider et al. 1980). When the tentacles are fully extended the phycobionts, located in digestive cells, form a monolayer just below the epidermis that maximally exposes the algae to light. When the tentacles contract, the phycobionts are repositioned, forming a layer 10–11 cells thick, effectively shading the innermost cells. Pearse (1974a, b) has shown that the phycozoan anemone *Anthopleura* positions itself in the light and exhibits varying degrees of contraction depending upon light intensity. That these behaviors are specific to the phycozoans is shown by the fact that corresponding aposymbiotic *Anthopleura* do not show these light-dependent phenomena. The flatworm phycozoan *Convoluta* orients toward the light and will settle preferentially on a light background (Keeble & Gamble 1907), and the green hydra exhibits a positive phototaxis (Pardy 1976). In the latter example positive phototaxis is exhibited by aposymbiotic hydra (Pardy 1976), and thus this characteristic seems to preadapt the phycozoan to seek a lighted environment.

There are other mechanisms that serve to regulate or optimize photosynthesis in phycozoans. Wethey and Porter (1976) have shown that among certain coral phycozoans the light saturation characteristics and kinetics of photosynthesis are modulated for depth. With increasing depth (and corresponding decreasing light intensity and a change in spectral quality) the photosynthetic apparatus undergoes a compensation optimizing the system for decreased light levels, so that deep-dwelling corals behave like shade plants.

Recent measurements on phycobionts isolated from green hydra and from *Aiptasia* have shown that photosynthesis by these bionts in vitro is highly inhibited by oxygen (Phipps & Pardy 1982). *Hydra* phycobionts in situ exhibited a 90% reduction of inhibition at ambient oxygen tensions over phycobionts analyzed in vitro under similar light and oxygen conditions. Furthermore, the maximum photosynthetic rate (P_{max}) was higher for phycobionts in situ. The physiological basis of these phenomena is presently unknown but may involve regulation of the relative concentrations of oxygen and carbon dioxide at the site of the phycobiont cells.

When oxygen measurements are used to determine photosynthetic rates of phycozoans it is invariably found that the rate of oxygen efflux exceeds influx. This implies net photosynthesis and indicates that oxygen production exceeds the total respiratory demand of the system. During maximum photosynthesis, bubbles of oxygen are observed floating off the surface of phycozoan corals (Crossland & Barnes 1977). Hence, it is not entirely surprising that some mechanism might exist to shield phycobionts from oxygen inhibition. It should be noted that during net photosynthesis the zoobiont experiences a reversal of the usual flow vectors for carbon dioxide and oxygen, that is, an *inward* flux of carbon dioxide and an *outward* flux of oxygen – a most untypical situation for an animal yet perfectly normal for a plant (and a phycozoan). In summary, phycozoans exhibit a combination of structural, biochemical, and behavioral preadaptations and other refinements that optimize photon capture, facilitate gas transport, and promote the functional integrity of the phycobionts.

In all phycozoans known, the symbiotic mode appears not to be optional. Even though aposymbiotic forms may be produced in the laboratory, it is doubtful if these ordinarily exist in nature. Hence, a key characteristic of phycozoans is their ability to self-perpetuate. Several mechanisms operate to assure the continual existence of phycozoans through successive generations of the zoobionts. Which particular mechanism is utilized by a given phycozoan depends upon the life history and reproductive mode of the phycozoan (sexual vs. asexual) and the location of the phycobionts in the system (intracellular vs. intercellular). Some mechanism must also exist for the controlled expansion and positioning of the phycobiont population as the zoobiont undergoes growth and development and experiences cell turnover of algal-bearing tissues. In phycozoans with endocellular phycobionts, for example, the green hydra, distribution of the algae among host daughter cells takes place during host cell division, or cytokinesis. With two possible exceptions (discussed later) direct cell-to-cell transfer of phycobionts is unknown. Distribution of phycobionts during host cell division accounts for the maintenance of algae in zoobiont tissues during growth and tissue turnover. During asexual reproduction of coelenterate phycozoans (corals, some sea anemones, jellyfish, hydra), buds are formed on the parent by cell division and tissue migration. Cells comprising bud primordia contain phycobionts originating from the parent's cells via cytokinesis. Hence developing buds are provided with a complement of algae as they grow and undergo morphogenesis. Development of the phycozoan, therefore, coincides with the development of the asexually produced zoobiont.

A prerequisite for the serial transfer of endocellular phycobionts through successive division cycles of host cells is the continued maintenance and replenishment of the algal population. If the growth rate of the phycobionts exceeds that of the zoobiont, the end result will be overpopulation leading to disintegration of the phycozoan. On the other hand, if the zoobiont's cells divide faster than the endocellular algae, the phycobionts will be diluted or even eliminated (Karakashian 1963, Pardy

1974). Although in some instances there appears to be an ongoing elimination of phycobionts (Taylor 1969), presumably to prevent overpopulation, the cellular-biochemical basis of regulation of host:symbiont cell ratio is unknown.

Fraser (1931) reported that eggs of *Myrionema*, a marine hydroid, contain algal symbionts. Eggs of green hydra have also been reported to contain symbionts (Brien & Renirs-Decoen 1950) though the observations have not been substantiated (Thorington & Margulis 1981). In both of these cases, cell-to-cell transfer of symbionts is postulated though the mechanism is unknown. With these possible exceptions, the gametes of zoobionts so far studied apparently do not contain algae (Trench 1980). Hence the provisional generalization emerges that, during sexual reproduction, the animal zygote is temporarily aposymbiotic and must reencounter free-existing symbionts. To effect a successful resynthesis at least two requirements must be met: (1) the zoobiont zygote must be receptive, and (2) it must encounter the appropriate algal cells. Establishment of the phycozoan usually occurs early in the postembryonic or larval phase of the zoobiont. Development of the animal must proceed to the point at which the larvae can feed because establishment of the phycozoan results from the ingestion of symbionts and subsequent uptake from the gut. Certain mechanisms operate to enhance a larva's chance of encountering appropriate algae. In some phycozoan corals and anemones, for instance, the planula larvae develop within the parent's enteron. Symbiotic algae are normally available here as a result of an extrusion process thought to be part of a population control mechanism (Steele 1976, Taylor 1969). The planulae thus become phycozoans prior to their release to the environment. In a similar manner the swimming larvae of phycozoan didemnids (colonial tunicates) acquire phycozoan status prior to dissemination (Eldredge 1965). In hydra, where the eggs develop externally, attached to the parent body wall, phycobionts are extruded from the maternal gastrodermal tissue, exit via the mouth, and come to adhere to the egg's surface (Thorington & Margulis 1981). Upon hatching, the larval hydra is thought to engulf some of these algae and becomes a phycozoan. In a similar manner the females of phycozoan flatworms (*Convoluta*) extrude mucus and phycobionts at the time of egg laying. This material, which coats the eggs, is ingested by larvae upon hatching, thus effecting the synthesis of the phycozoan. There is some evidence that *Convoluta* eggs also possess chemotactic factors that attract algae (Holligan & Gooday 1975).

The final establishment of a stable phycozoan resulting from ingestion of potential algal symbionts, either by larvae or adult zoobionts, is not an event but appears rather to be a sequence of processes involving uptake of the algae, positioning of the algae within the cells or appropriate tissues, and integration of the algal and animal physiology (Smith

1979). A variety of studies have shown that zoobionts are not indiscriminate but exhibit a degree of selection or preference for certain algal species or strains. For example, Provasoli et al. (1968) have shown that *Convoluta* will associate with several different platymonads but these are eliminated when the worm is presented with the "preferred" species. A similar finding has been reported for *P. bursaria* (Karakashian 1963). Trench et al. (1981) have shown that the polypoid stage of the jellyfish *Cassiopiea* is aposymbiotic and must reacquire algae from the environment. From a heterogeneous assemblage of phytoplankters, the polyps in some way "selected" only one characteristic species. Of additional interest is that if they do not acquire symbionts, polyps cannot strobilate (produce new jellyfish). Thus only as a phycozoan can *Cassiopiea* complete its life cycle.

Experiments with green hydra have suggested the existence of a host-symbiont recognition system in green hydra. *Hydra* aposymbionts reject most algae, preferring symbionts originating from green hydra (Pardy & Muscatine 1973). Moreover, the reacquisition process, which is inhibited by treating symbionts with plant lectin (Meints & Pardy 1980), may involve specific surface antigens on the algae (Pool 1979) and may require the mobilization of microtubules (Cooper & Margulis 1978, Fracek & Margulus 1979). An elaboration of some of these aspects, along with a theoretical discussion can be found in Jolley and Smith (1980). The perpetuation of phycozoans is apparently not left to chance but involves processes (in some cases, highly specific mechanisms) to effect successful resynthesis.

If the concept of phycozoans can be accepted, then qualitative and quantitative statements about the ecological niche defined by these organisms should be possible. With the exception of work on some reef-forming corals (Muscatine & Porter 1977) our understanding of the phycozoan niche is undeveloped. Virtually nothing is known of the genetics of phycozoans. Moreover, we can only speculate on how the forces of evolution operate on these organisms. We are just beginning to appreciate how they are stabilized and perpetuated.

I have tried not to be dogmatic about what is or is not a phycozoan. I hope more precisely defined criteria will emerge as our knowledge increases. At one extreme there is the case where algae live in the hairs of the polar bear (Lewin et al. 1981). Does this make the polar bear a phycozoan? Probably not. At the opposite pole (no pun intended) there is *Euglena*. Is it a phycozoan? Maybe. In conclusion, I have tried to show that what we consider to be an association of separate organisms (algae living in invertebrates) may actually be a whole organism. This organism retains many properties of its component bionts but also has features that assure its stability and perpetuation. I suggest that these algal-invertebrate symbioses be called phycozoans to emphasize their plant-animal nature and their uniqueness. The study of phycozoans is, therefore, phycozoology, and those who study them are phycozoologists.

References

Boyle, J. E., & D. C. Smith (1975). Biochemical interaction between the symbionts of *Convoluta roscoffensis*. *Proc. Royal Soc. Lond.* Series B. **189**, 121–35.

Brien, P., & M. Renirs-Decoen (1950). Etude d'*Hydra viridis* (Linnaeus): La bastogenese, la spermatogenese, l'ovogenese. *Ann. Royal Zoo. Soc. Belgium* **81**, 33–110.

Cernichiari, E., L. Muscatine, & D. C. Smith (1969). Maltose excretion by symbiotic algae of *Hydra viridis*. *Proc. Royal Soc. Lond.* Series B. **173**, 557–76.

Cook, C. B. (1980). The infection of invertebrates by algae. *In: Cellular Interactions in Symbiosis and Parasitism*, ed. C. B. Cook, F. W. Pappas, and E. D. Rudolph, pp. 47–73. Columbus: Ohio State University Press.

Cooper, C. G., & L. Margulis (1978). Delay in migration of symbiotic algae in *Hydra viridis* by inhibitors of microtubule protein polymerization. *Cytobios* **19**, 7–19.

Crossland, C. J., & D. J. Barnes (1977). Gas-exchange studies with the staghorn coral *Acropora acuminata* and its zooxanthellae. *Mar. Bio.* **40**, 185–94.

Dorey, A. E. (1965). The organization and replacement of the epidermis in acoelous tubellarians. *Quar. J. Micros. Sci.* **106**, 147–72.

Eldredge, L. G. (1965). A taxonomic review of Indo-Pacific didemnid ascidians and descriptions of twenty-three Central Pacific species. *Micronesia* **2**, 161–261.

Fogg, G. E. (1962). Extracellular products. *In: Physiology and Biochemistry of Algae*, ed. R. A. Lewin, pp. 475–89. New York: Academic Press.

Fracek, S., & L. Margulis (1979). Colchicine, nocodazole and trifluralin: different effects of microtubule polymerization inhibitors on the uptake and migration of endosymbiotic algae in *Hydra viridis*. *Cytobios* **25**, 7–15.

Fraser, E. A. (1931). Observations on the life history and development of the hydroid *Myrionema ambionense*. *Sci. Rep. Great Barrier Reef Expedition* **3**, 135–44.

Glider, W. V., D. W. Phipps, Jr., & R. L. Pardy (1980). Localization of symbiotic dinoflagellate cells within the tentacle tissue of *Aiptasia pallida* (Coelenterata, Anthozoa). *Trans. Amer. Micros. Soc.* **99**, 426–38.

Holligan, P. M., & G. W. Gooday (1975). Symbiosis in *Convoluta*. *In: Symbiosis. Symp. Soc. Exper. Bio.* **32**, 205–7.

Hutchinson, G. E. (1959). Homage to Santa Rosalina or Why are there so many kinds of animals? *Amer. Nat.* **93**, 145–59.

Jolley, E., & D. C. Smith (1980). The green hydra symbiosis. II. The biology of the establishment of the association. *Proc. Royal Soc. Lond.* Series B. **207**, 311–33.

Karakashian, S. J. (1963). Growth of *Paramecium bursaria* as influenced by the presence of algal symbionts. *Phys. Zoo.* **36**, 52–68.

Keeble, J., & J. W. Gamble (1907). The origin and nature of the green cells of *Convoluta roscoffensis*. *Quar. J. Micros. Sci.* **51**, 167–219.

Lewin, R., P. A. Farnworth, & G. Yamanka (1981). Algae of green polar bears. *Phycologia* **20**, 303–14.

Meints, R., & R. L. Pardy (1980). Quantitative demonstration of cell surface involvement in a plant-animal symbiosis: Lectin inhibition of reassociation. *J. Cell Sci.* **43**, 239–57.

Muscatine, L. (1965). Symbiosis of hydra and algae. III. Extracellular products of the algae. *Comp. Biochem. Phys.* **16**, 77–92.

– (1967). Glycerol excretion by symbiotic algae from corals and *Tridacna* and its control by the host. *Science* **156**, 516–9.

– (1974). Endosymbiosis of cnidarians and algae. *In: Coelenterate Biology*, ed. L. Muscatine & H. M. Lenhoff, pp. 359–95. New York: Academic Press.

Muscatine, L., & E. Cernichiari (1969). Assimilation of phytosynthetic products of zooxanthellae by a reef coral. *Biol. Bull.* **137**, 506–23.

Muscatine, L., & H. M. Lenhoff (1965). Symbiosis of hydra and algae. II. Effects of limited food and starvation on growth of symbiotic and aposymbiotic hydra. *Biol. Bull.* **129**, 316–28.

Muscatine, L., & J. W. Porter (1977). Reef corals: Mutualistic symbioses adapted to nutrient-poor environments. *BioScience* **27**, 454–60.

Muscatine, L., S. J. Karakashian, & J. W. Karakashian (1967). Soluble extracellular products of algae symbiotic with a ciliate, a sponge and a mutant hydra. *Comp. Biochem. Phys.* **20**, 1–12.

Muscatine, L., C. B. Cook, R. L. Pardy, & R. R. Pool (1975). Uptake, recognition and maintenance of symbiotic chlorella by *Hydra viridis*. *Symp. Soc. Exper. Bio.* **29**, 175–203.

Oschman, J. L. (1966). Development of the symbiosis of *Convoluta roscoffensis* Groff and *Platymonas* sp. *J. Phycol.* **2**, 105–11.

Pardy, R. L. (1974). Some factors affecting the growth and distribution of the algal endosymbionts of *Hydra viridis*. *Biol. Bull.* **147**, 105–18.

– (1976). Aspects of light in the biology of green hydra. *In: Coelenterate Ecology and Behavior*, ed. G. O. Mackie, pp. 401–7. New York: Plenum Press.

– (1980). Symbiotic algae and ^{14}C incorporation in the freshwater clam, *Anodonta*. *Biol. Bull.* **158**, 349–55.

Pardy, R. L., & L. Muscatine (1973). Recognition of symbiotic algae by *Hydra viridis*. A quantitative study of the uptake of living algae by aposymbiotic *H. viridis*. *Biol. Bull.* **145**, 565–79.

Pardy, R. L., & B. White (1977). Metabolic relationships between green hydra and its symbiotic algae. *Biol. Bull.* **153**, 228–36.

Pearse, V. B. (1974a). Modification of sea anemone behavior by symbiotic zooxanthellae: Phototaxis. *Biol. Bull.* **147**, 630–40.

– (1974b). Modification of sea anemone behavior by symbiotic zooxanthellae: Expansion and contraction. *Biol. Bull.* **147**, 641–51.

Phipps, D. W., Jr., & R. L. Pardy (1982). Host enhancement of symbiont photosynthesis in the hydra-algae symbiosis. *Biol. Bull.* **162**, 83–94.

Pool, R. R. (1979). The role of antigenetic determinants in the recognition of potential algal symbionts by cells of *Chlorohydra*. *J. Cell Sci.* **35**, 367–79.

Porter, J. W. (1976). Autotrophy, heterotrophy and resource partitioning in Caribbean reef-building corals. *Amer. Nat.* **110**, 731–42.

Provasoli, L., T. Yamasu, & I. Manton (1968). Experiments on the resynthesis of symbiosis in *Convoluta roscoffensis* with different flagellate cultures. *J. Mar. Bio. Ass. U. K.* **48**, 465–79.

Smith, D. C. (1974). Transport from symbiotic algae and chloroplasts to animal hosts. *Symp. Soc. Exp. Bio.* **28**, 473–508.

– (1979). From extracellular to intracellular: the establishment of a symbiosis. *Proc. Royal Soc. Lond.* Series B. **204**, 1–16.

Smith, D. C., L. Muscatine, & D. H. Lewis (1969). Carbohydrate movement from autotrophs to heterotrophs in parasitic and mutualistic symbiosis. *Biol. Rev.* 44, 17–90.

Starr, M. P. (1975). A generalized scheme for classifying organismic associations. *In: Symbiosis*, ed. D. H. Jennings and D. L. Lee, pp. 1–20. Cambridge: Cambridge University Press.

Steele, R. D. (1976). Light intensity as a factor in the regulation of the density of symbiotic zooxanthellae in *Aiptasia tagetes* (Coelenterata, Anthozoa). *J. Zoo.* 179, 389–405.

Taylor, D. L. (1969). On the regulation and maintenance of algal numbers in zooxanthellae-coelenterate symbiosis with a note on the nutritional relationship in *Anemonia sulcata*. *J. Mar. Bio. Ass. U. K.* 49, 1057–65.

Thorington, G., & L. Margulis (1981). Transmission of the algal and bacterial symbionts of green hydra through the host sexual cycle. *In: Symbiosis and Evolution of Cell Organelles*, ed. W. Schwemmler & H. E. A. Schenk, vol. 1, pp. 175–222. Berlin: Walter de Gruyter.

Trench, R. K. (1971). The physiology and biochemistry of zooxanthellae symbiotic with marine coelenterates. II. Liberation of fixed ^{14}C by zooxanthellae *in vitro*. *Proc. Royal Soc. Lond.* Series B. 177, 237–50.

– (1975). Of "leaves that crawl." Functional chloroplasts in animal cells. *Symp. Soc. Exp. Bio.* 29, 229–65.

– (1979a). The cell biology of plant-animal symbiosis. *Ann. Rev. Plant Phys.* 30, 485–531.

– (1980). Integrative mechanisms in mutualistic endosymbioses, *Ann. Coll. Biol. Sci. Colloq.*, 5th. Columbus: Ohio State University Press.

Trench, R. K., N. J. Colley, & W. K. Fitt (1981). Recognition phenomena in symbioses between marine invertebrates and "zooxanthellae;" uptake, sequestration and persistence. *Ber. Deutsch. Bot. Ges.* 94, 529–45.

Wethey, D. S., & J. W. Porter (1976). Sun and shade differences in productivity of reef corals. *Nature* 262, 281–2.

Whitney, D. D. (1907). The artificial removal of the green bodies of *Hydra viridis*. *Biol. Bull.* 13, 291–9.

2

The coral-algal symbiosis

DENNIS L. TAYLOR

Center for Environmental and Estuarine Studies
University of Maryland
Cambridge, MD 21613

> The maintenance of zooxanthellae within the corals and in all other coelenterates and certain other reef organisms, represents an imprisonment of the phytoplankton. These plants obtain the raw materials for carbohydrate and protein synthesis direct from the animals in which they live.
>
> C. M. Yonge (1958)

Dinoflagellates are widely distributed in marine, estuarine, and freshwater environments, while motile acellular forms are frequently important constituents of phytoplankton populations. They are ancient, enigmatic organisms, exhibiting unique structural, physiological, and biochemical features that obscure their phylogeny and offer little relief to those desiring a positive systematic placement. In marine environments, they readily participate in a variety of symbiotic consortia, serving as both hosts and symbionts. As hosts, they establish unique cellular chimeras (Tomas & Cox, 1973; Jeffrey & Vesk, 1976; Gibbs, 1981) that offer opportunities to test the concept of intergenomic cooperation (Bogorad, 1975) and to examine the evolutionary consequences of the symbiotic condition. As symbionts, they have emerged as a potent biogeochemical force, serving to structure and stabilize shallow-water tropical marine ecosystems both physically and biologically.

It is the latter role of symbiont that is most widely recognized. These dinoflagellates are the "zooxanthellae" of both modern and classic reference. They are the "imprisoned phytoplankton" of Yonge (1958), growing as a unialgal bloom sustained by symbiotically regulated mechanisms of nutrient acquisition, conservation, and cycling. These mechanisms are the result of mutual adaptations between dinoflagellate symbionts and their invertebrate hosts. They serve to strengthen and stabilize the functional symbiotic unit, while obscuring the discrete physiological and biochemical attributes of the two symbiotic partners. It is both difficult and unprofitable to discuss the nutritional physiology and ecology of symbiotic dinoflagellates outside the context of the functional relationship they have with their host. In culture, dinoflagellate symbionts differ little from their free-living relatives, suggesting that symbiotic attributes are manifested largely through interactions with host species. However, it is clear that symbionts have a predisposition for symbiosis that has yet to be rigorously described. Comparative in vitro and in vivo studies can contribute to this description and assist in the description of the fundamental nature of symbiotic interactions (Taylor, 1980).

The major biological questions and principal experimental themes that lead toward an understanding of the nutritional physiology and ecology of algal symbionts in general were first elaborated by Keeble (1910). Nutritional interdependency, born of the need to conserve and recycle nutrients and energy in impoverished habitats, is central to the success of symbiotic relationships involving dinoflagellates. The frequency of their occurrence is positively correlated with nutrient-poor environments (Muscatine & Porter, 1977), where symbiosis offers the ultimate practical solution. Characteristically, these relationships yield improved primary production and a vigorous benthic carbon metabolism (Taylor, 1973a, b; Sournia, 1977).

Coral reefs are perhaps the best-known manifestation of a symbiotically structured ecosystem. Their growth and evolution since the Triassic rest to a considerable degree upon the perpetuation of an endosymbiotic relationship between the hermatypic corals and dinoflagellates, notably *Gymnodinium microadriaticum* (Freudenthal) and its closely allied subspecies and varieties. These relationships permit the exploitation of nutrient-impoverished seas and establish the form and architecture of habitats utilized by myriad other organisms. Most important, they establish major zones of high primary production and, through vaguely understood mechanisms linking dinoflagellate photosynthesis with calcification (Chalker & Taylor, 1975; Chalker, 1976), they participate in a major biogeochemical process affecting the overall calcium budget of the oceans (Smith, 1978).

Organisms and ecosystems do not evolve without need or opportunity. Symbiosis in reef corals provides an opportunity for responsive coevolution with the environment. Together, the nutritional and depositional consequences of these associations permit both the invasion and conquest of new environments and their stabilization through the establishment of a persistent physical structure that is itself an evolving habitat (Taylor, 1981). The following discussion examines both the nature and the consequences of the various cellular interactions that take place in dinoflagellate-coral symbioses. As will be seen, these may be genetic, chemical, and physiological in nature.

What's in a name?

Hermatypic (reef-building) corals are defined by their symbiosis with dinoflagellates. The association provides investigators with a useful, subjective characteristic that clearly distinguishes them from the solitary, ahermatypic species. Undoubtedly, symbiosis has been a significant factor in the evolution of hermatypic species. As a result, the distinction between hermatypes and ahermatypes generally involves significant systematic distance, although it may also occasionally involve species of the same genus (e.g., *Madracis*) or a single, facultatively symbiotic species.

Within associations involving hermatypic species, dinoflagellate symbionts tend to exhibit a strikingly uniform morphology regardless of host origin or geographic location. Descriptions of these symbionts based on observations of vegetative stages in situ, and examinations of motile stages without benefit of detailed studies in culture, are flawed by this apparent uniformity. Use of the older (i.e., nineteenth-century) literature for species identification is particularly difficult, and considerable confusion and disagreement among modern workers exist as a result. Where multiple symbioses involving more than one alga in a given host exist, critical identification and existence of type descriptions are extremely doubtful. Their use for the resurrection of genera and species is tenuous at best.

The most common dinoflagellate species recognized in associations involving corals is *G. microadriaticum*. Use of the genus *Gymnodinium* is an obvious statement of my preference. It is based upon a systematic viewpoint that emphasizes the *total* morphology of the motile stage as having primary systematic importance for these algae (Taylor, 1971) and follows a suggestion first put forth by Ball (1968). Recent literature on this organism would suggest that my premise in this matter is more narrowly restricted to "the position of the transverse groove on the motile cell" and that I regard this as "the only feature of systematic value in determining generic assignment" (Loeblich & Sherley, 1979). An examination of the literature cited by Loeblich and Sherley (Taylor, 1971, 1973b) should be sufficient to correct any errors of understanding. Other authors have other nomenclatural preferences. Witness the present volume. The most common generic alternatives in current use are *Symbiodinium microadriaticum* and *Zooxanthella microadriatica*. I would reject *Symbiodinium* on the grounds that its description is based upon habit rather than morphology and is therefore not founded on generally accepted principles applied to the protista (Ball, 1968; Taylor, 1971). Similarly, I do not favor *Zooxanthella* for the following reasons. It is uncertain whether more than one symbiont exists in association with *Collozoum inerme*, the type host described by Brandt (1881). My examination of this host collected from the type locale in 1968 revealed a symbiont with obvious amphidinium morphology in the motile stage (Taylor, 1974a). Hollande and Carré (1975) also examined *C. inerme* from a different locale, but found symbionts with a gymnodiniumlike morphology. Clearly these observational differences need to be resolved. If, as the present evidence indicates, *C. inerme* is symbiotic with more than one dinoflagellate, then we have no way of positively identifying the type organism from Brandt's (1881) original description and illustration. The fact that zooxanthella is also a term in vulgar usage referring generally to yellow-brown pigmented symbionts of dinoflagellate, cryptomonad, chrysomonad, and diatom affinities only further detracts from its desirability as a formal genus.

Coral-dinoflagellate symbiosis emerged prominently in the Triassic, but could well have originated much earlier, possibly during the Ordovician. In general, the Dinophyceae are an extremely ancient assemblage, with fossil remains in the Cambrian. As a group, they are cosmopolitan in their present distribution. This characteristic is shared by symbiotic species and may be explained in part through present knowledge of continental drift, notably by an analysis of the position of the continental masses in the Triassic, and their subsequent movements following the breakup of the supercontinents Laurasia and Gondwana. Within the milieu of a given coral host species, genetic isolation may occur and with it the potential for the evolution of different symbionts adapted specifically to that host. Conversely, one might also argue that the uniformity of the symbiotic state would mitigate strongly against any symbiont evolution and that the present morphological uniformity of dinoflagellate symbionts is an expression of that fact. The truth appears to lie somewhere between the concept of a single pandemic species and that of a single species for each host (Taylor, 1973b, 1974a). Present evidence obtained from electrophoretic studies of isozymes reveals the existence of several races or varieties that may be characteristically associated with a given host (Schoenberg & Trench, 1980a–c). Whether such differences are adequate for the separation of individual species as suggested (Loeblich & Sherley, 1979) remains to be seen. The expression of a given host's or symbiont's genetic suite should be viewed in the context of the functional unit. Within that context gene expression may be significantly altered or suppressed (Taylor, 1981), and traits such as isozyme patterns may be more a reflection of gene expression within the total gene pool of the functional unit than a stable genetic characteristic of the symbiont per se. At present we do not know the extent to which gene expression is altered by the cellular propinquity of the symbiotic state, although there is evidence that such effects can occur in other cellular systems (Kollar & Fisher, 1980). DNA hybridization studies and an analysis of DNA homologies would provide a useful assessment of genetic separation among strains of *G. microadriaticum* that would serve to resolve many of the current species-related questions.

Systematic onanism aside, there are important practical reasons for agreeing upon a name for what is clearly the most prominent coral symbiont, since it will serve to reconcile much of the existing literature. However, it is fortunate that concern over a name does not overshadow our ability to gather data on the nutritional physiology and ecology of associations between *G. microadriaticum* and corals. As we shall see, that which I call *G. microadriaticum* by any other name will have the same cellular interactions. Our interest in these cellular interactions and their ecological consequences is by far our ultimate reason for studying them.

The acquisition, exchange, and cycling of nutrients and energy

The nutritional consequences of algal-invertebrate symbiosis have been reviewed extensively with more than adequate attention directed toward the specifics of coral-dinoflagellate associations (Muscatine, 1973, 1980; Taylor, 1973a, b; Trench, 1979). Another elaborate review at this time would be merely repetitive and unnecessary. I will therefore merely highlight the important features that define the successful functional unit we observe in nature. Before proceeding, it is worth noting two aspects of the recent literature that are revealed in reviews of the subject. The first is that although research in the field has contributed to increasingly finer resolution of data, the approach and perspective of this research have remained static for some time. There is a clear need for new and innovative approaches to the study of symbiotic systems in general. An ecological perspective that views the cell as a habitat is one avenue that can be effectively exploited in future studies (Smith, 1979; Gooday & Doonan, 1980). Another perspective views the problems of cellular, molecular, and genetic integration in symbiotic systems as essentially the problems of multicellularity and cellular evolution. Both may be equally fruitful. The second aspect is that the focus of research has emphasized investigations that either inventory materials and resources available to the intact association or demonstrate the existence of specific transport and uptake phenomena that are required to acquire these materials and resources. In short, existing data are a catalog of what enters via uptake and what leaves via excretion. This catalog does not illuminate the internal workings of the functional unit to any significant degree and fails as an explanation of why, in a symbiosis, the sum is greater than the two parts. It merely points instead to a significant philosophical and experimental deficiency that needs to be addressed.

Acquisition and exchange

The functional unit of coral-dinoflagellate symbiosis is polytrophic, having the ability to access a complete suite of nutritional resources found in reef environments (Muscatine & Porter, 1977; McCloskey et al., 1978; Muscatine, 1980). These would include dissolved and particulate organic materials, living prey, and photosynthesis. The individual importance of each of these must be assessed in terms of the total nutritional resources available to the symbiosis. Consideration must also be given to the ability of hosts and symbionts to fully utilize these opportunities when acquiring materials and energy. To this end, several useful speculative models have been developed for the pathways of nitrogen and carbon in symbiotic associations (Lewis & Smith, 1971; Muscatine & Porter, 1977; Smith, 1979). These are currently being used to develop resource and utilization budgets for symbioses involving reef corals.

Dissolved and particulate organic matter. Reef corals have traditionally been characterized as one of the most highly specialized carnivores in the animal kingdom (Yonge, 1930), a viewpoint that has generally overshadowed the significance of osmotrophy and suspension feeding by these organisms. Studies of the utilization of dissolved organic materials are scant (Stephens, 1962), and it is difficult to establish the significance of this activity as a phenomenon external to the translocation of dissolved materials in the functional symbiotic unit. It is likely that uptake from the surrounding milieu is merely a reflection of the predisposition of the symbiosis to the exchange of dissolved organic compounds between partners and that the internal resource will always exceed the requirement for an external supply. There is also the question of whether the reef habitat actually has a significant level of dissolved organic material that would be of qualitative and quantitative importance to the coral-dinoflagellate association.

Dinoflagellate symbionts inhabit a rich organic milieu that exists within host tissues and cells. This circumstance suggests that heterotrophy may play an important nutritional role for the alga and that it may serve a broader function as a mechanism of nutrient conservation in the intact symbiosis. Although it may be regarded as a paradox, present evidence indicates that significant heterotrophy is unlikely. In culture, *G. microadriaticum* exhibits only limited heterotrophic abilities (Taylor, 1974a, in press). It may be shown to utilize a variety of amino acids as sources for growth and metabolism and for the direct synthesis of protein. Studies of carbohydrates indicate that sugars are metabolized, but they cannot support the growth of symbionts. It is doubtful that these abilities are significant for symbionts in vivo. Alternative nitrogen sources such as ammonia, nitrate, and urea compete effectively with amino acids and are utilized preferentially by symbionts in competitive experiments (Taylor, unpublished). Other compounds, notably sulfate, taurine, cysteine, and methionine, are readily taken up and assimilated as sources of sulfur (Deane & O'Brien, 1981a). As sources of trace nutrients such activities may have significance for the alga.

Reef environments are known to contain large amounts of suspended particulate material derived from plant and animal residues (Johannes et al., 1970; Glynn, 1973), which may be a superior food resource due to bacterial colonization of the particles. Similarly, mucus flocs present in the water column may be an important reservoir of carbon for the symbiosis (Ducklow & Mitchell, 1979a, b; Rublee et al., 1980). Several authors have examined the utilization of these particulate materials. DiSalvo (1971) and Sorokin (1973) provide excellent summaries of bacterial utilization. Other authors have examined the mechanism of particle entrapment. Typically, corals will form mucus nets that greatly increase surface areas and facilitate capture over considerable distances from the colony (Roushdy & Hansen, 1961; Lewis, 1976; Lewis & Price, 1975). To date, there is no substantive study of particle utilization that provides reasonable estimates of the contribution this resource makes to the total carbon budget of the coral symbiosis. It is likely that this contribution is highly variable, depending ultimately upon the individual species and on the nutritional emphasis in effect for any given individual. The available evidence suggests that the amount of particulate material available exceeds the demand. Thus the limiting factor is demand, not supply.

Living prey. The contribution of zooplankton to the nutritional budget of coral-dinoflagellate associations has been summarized recently (Muscatine & Porter, 1977). A key issue has always been the need to determine the magnitude and nutritional suitability of reef zooplankton and to relate this to the impressive prey capture potential of the coral colony (Yonge, 1930) and the actual metabolic demand placed on the zooplankton resource in the total nutritional budget of the functional unit. Given the wealth of available data examined by Muscatine and Porter (1977), they conclude that zooplankton feeding does not contribute a majority of either the calories or the carbon required by the symbiosis. There is general agreement on this point among workers in the field. However, the question remains with respect to how one may resolve the existence of a powerful and efficient prey capture mechanism with the relative insignificance of available prey in the materials and energy budget of coral-dinoflagellate symbiosis. Recent work by Lasker (1981) on *Montastrea cavernosa* provides some insights into this problem. Given the importance of symbiont photosynthesis to a coral's nutritional balance, it is possible to speculate that the machinery of prey capture (polyps and tentacles) has been retained in the course of the evolution of these associations, but it has been functionally diverted to serve instead as a means of increasing surface area for the capture of photons instead of particles and zooplankton. The study of phenotypic variation in *M. annularis* offers important clues to how this may occur (Lasker, 1981).

Photosynthesis. Within the functional unit of a reef coral, symbiont photosynthesis provides the principal source of reduced carbon in the majority of associations. Yonge's (1958) observation that the symbionts of corals are imprisoned phytoplankton is a valid and useful perspective that relates not only to the growth and nutrition of hosts and symbionts participating in coral-dinoflagellate associations, but also to their ecology and the systematics of the dinoflagellate itself.

Photosynthetic pathways in *G. microadriaticum* have been examined tentatively by several workers. Our most detailed knowledge relates to the nature of photosynthetic products in vitro and in vivo (Muscatine, 1980; Taylor, in press). Carbon fixation is generally believed to follow the C_3 pathway, but there is increasing evidence that, like other dinoflagellates, *G. microadriaticum* exhibits a mixed C_3–C_4 metabolism (Trench, 1979; Taylor, in press). The analysis of photosynthetic products produced reveals a substantial range of organic compounds (Schmitz & Kremer, 1977). Comparative studies in vitro and in vivo are important here as a means of establishing products resulting from secondary metabolism by the host. Such materials are readily translocated to the host, where they are metabolized to yield energy and build tissue. In such a system, the host functions as a primary consumer if the symbiosis is viewed ecologically. Alternatively one may view the functional unit as analogous with the multicellular condition of a higher plant where a substantial portion of the photosynthetic output is directed toward the construction of what Meyers (1980) has referred to as "skeletal crud." In a functional sense, these "imprisoned phytoplankton" have made the transition from a unicellular existence where photosynthetic production emphasizes cell machinery to a multicellular existence emphasizing structure and the division of labor. Only through symbiosis can this evolutionary leap be accomplished so swiftly.

O_2 production and consumption by *G. microadriaticum* in vitro has been examined (Burris, 1977) and is currently under investigation in my laboratory using mass spectrometry and $^{32}O_2$ and $^{36}O_2$. The apparatus and methodology are as described by Bunt (1969). The early data from these investigations indicate a level of photorespiration that is at times comparable to that reported for *Chlorella* (Bunt & Heeb, 1971). This activity has not been examined in vivo. Although it is reasonable to presume that photorespiration may be unimportant in vivo due to unfavorable O_2 and HCO_3^- concentrations (Muscatine, 1980), photorespiration may have an effect upon the quantitative analysis of photosynthetic products in vitro and would thus affect our ability to calibrate in vitro and in vivo studies of production for the analysis of intercellular exchange.

Primary production. Within the benthic ecosystem of the coral reef, algal-invertebrate symbiosis dominates. Among these associations, those involving corals and dinoflagellates are most often in the majority. By its very nature, the coral colony with its associated symbionts represents an exceptionally large veneer of photosynthetically active tissue covering much of the surface area of the reef above 100 m. Because of this, there is substantial interest in the hermatypic corals as primary producers. Muscatine and his colleagues have reviewed the contribution that *G. microadriaticum* makes toward the total productivity of coral reefs (Muscatine, 1980; Muscatine et al., 1981), paying particular attention to the inherent difficulties in assessing the relative values of net photosynthesis, gross photosynthesis, coral respiration, and symbiont respiration. The equation they have developed is useful as a focus for a renewed effort in field determinations. Coral reefs are generally regarded as one of the most productive of the earth's ecosystems. Exactly how productive is a question that awaits a reliably quantified answer.

Nitrogen and phosphorus. Symbiosis offers an effective strategy to ameliorate the impact of limiting nutrients on primary production and growth. In aquatic environments, nitrogen appears as the key limiting nutrient, and it is likely that many algal-invertebrate associations are reinforced and stabilized by the "universal nitrogen hunger" of the partners (Keeble, 1910). A cardinal feature of such associations is their ability to detect and remove nitrate and ammonia from the medium at extremely low concentrations (D'Elia & Webb, 1977). This is true regardless of whether one is dealing with a natural association or one created experimentally (Taylor, 1978a). The principle has been recognized for some time (Yonge, 1936), and it is central to the success of symbiotically based ecosystems in nutrient-poor waters. Several potential mechanisms for nitrogen conservation exist within the body of compounds produced and translocated in the symbiosis. These depend upon (1) the abilities of symbionts to grow on a wide range of nitrogen sources (Taylor, 1975), most notably ammonia and urea; (2) translocation of potential carrier compounds such as alanine, glutamate, and aspartate (Lewis & Smith, 1971; Boyle & Smith, 1975); and (3) the ability of hosts to metabolize these carriers, thereby securing them as a constant nitrogen resource. Nitrogen can thus be conserved via the mutual excretory and assimilatory activities of hosts and symbionts. Although these ideas have not been rigorously tested, several hypothetical pathways have been discussed by Lewis and Smith (1971) and are further elaborated by Muscatine and Porter (1977). It is interesting to note in this connection that the shorter wavelengths of light that prevail in most habitats where symbioses occur tend to favor production of the previously noted amino acids. Light quality may therefore serve to strengthen and reinforce internal cycling of nitrogen.

Additional information on the nitrogen metabolism of dinoflagellate symbionts has been obtained from studies of *G. microadriaticum* using

^{15}N analysis to determine the uptake and utilization of $^{15}NH_4Cl$, $Na^{15}NO_3$, and $(^{15}NH_2)_2CO$ (Summons & Osmond, 1981). Assimilation in all cases is clearly light dependent and is most rapid in the case of NH_3, confirming earlier reports (Taylor, 1978b). The distribution of label in cell products is consistent with metabolism occurring via the enzymes glutamine synthetase and glutamate synthetase rather than via the alternative glutamate dehydrogenase pathway. Although these data are preliminary in nature, and the suggested pathway requires further experimental and kinetic analysis, they are important as the first real insight into the pathways of nitrogen utilization in a dinoflagellate symbiont.

In vitro studies of continuous cultures where nitrogen is held as the limiting nutrient have shown clear differences in the preferences individual symbionts may have for the form of nitrogen presented (Taylor, 1980). In competitive experiments involving paired symbionts, symbiont success is directly attributable to individual kinetic advantages, which may serve in nature as a primary factor in symbiont selection in a given host. Continued analysis of various strains of *G. microadriaticum* may contribute to an understanding of host selectivity in reef ecosystems where a variety of motile, infective symbionts are commonly available to a range of coral species.

Phosphorus is an essential element for cell growth, and it has been shown to be particularly important for the growth of dinoflagellate symbionts. Studies of reef corals by D'Elia (1977) demonstrate that the removal of phosphate from seawater is a function of the presence of the dinoflagellate symbiont and that removal can be accomplished at low ambient concentrations. Patterns of uptake are complex and appear to be largely dependent upon photosynthetic energy, although uptake may proceed in the dark by passive diffusion or facilitated diffusion (Deane & O'Brien, 1981b). Clearly, the presence of symbionts provides the host with an important mechanism for the acquisition of phosphate. Within the nutrient-deficient environment of a coral reef, this may be a significant attribute.

Cycling

One of the characteristic features of any symbiotic association is the movement (translocation) of substances between host and symbiont in one or both directions (Smith, 1979). More precisely, this characteristic is a property of altered permeabilities. Translocation of nutrients and energy between the alga and the host has been studied almost exclusively in coelenterates (Smith et al., 1969; Muscatine, 1980). Release of compounds by *G. microadriaticum* in vitro is stimulated by the presence of host tissues or extracts. Presumably this occurs in vivo as well although the mechanism remains a mystery 16 years after its first demonstration (Muscatine, 1967). Known compounds that could satisfy the nutritional requirements of the host represent low molecular weight, soluble materials that can readily access host metabolic pathways and satisfy requirements for energy and biosynthesis (see Smith et al., 1969; Muscatine, 1980). They are, in one sense, a biological reservoir of reduced carbon and essential nutrients. Other materials conserved by the alga and translocated to the host include nitrogen and phosphorus. As noted, comparatively little is known about their acquisition and cycling in symbiotic systems. Demonstration of alanine movement in substrate inhibition experiments provides proof of the translocation of nitrogen in this form (Lewis & Smith, 1971). Subsequent incorporation into protein has been postulated (Muscatine & Cernichiari, 1969) and confirmed in aposymbiotic hosts (Taylor, 1974b). Organic phosphates may function similarly to conserve and recycle phosphorus.

A dominant concern in coral physiology has been with the potential energetic significance to the host of carbon obtained via translocation (Muscatine, 1980). Given the shortfall in utilization of other resources (e.g., soluble and particulate organic carbon, prey capture), production and translocation become central to our consideration of reef corals as polytrophs emphasizing either autotrophy or heterotrophy. Accurate measurement of inputs and the development of resource and energy budgets based upon these data, coupled with realistic interpretations of classic photosynthesis : respiration ratios, offer a studied approach to the development of an accurate assessment of the nutritional emphasis of the intact functional unit.

Symbiont-enhanced skeletogenesis in hermatypic corals

The nutritional benefits that hermatypic corals gain from their symbiosis with *G. microadriaticum* are only one aspect of a relationship that is essential to their success as the foundation species in coral reef ecosystems. The rapid deposition of $CaCO_3$ during skeletogenesis is equally important, since it is the critical factor affecting the structure and architecture of reef habitats. It has been recognized for some time that the presence of algal symbionts in the tissues of hermatypic species stimulates the rate of $CaCO_3$ deposition and that this is most apparent in the light (Kawaguti & Sakumoto, 1948). Subsequent studies using radioisotopes have confirmed this, while demonstrating an unequivocal relationship between the rate of symbiont photosynthesis and the rate of skeletogenesis (Goreau & Goreau, 1959; Pearse & Muscatine, 1971; Vandermeulen et al., 1972; Chalker & Taylor, 1975). As with other aspects of this symbiosis, our knowledge is limited to the external manifestations of the phenomenon. There is considerable speculation as to the metabolic pathways and systems of transport involved internally, but relatively little is known about the actual inner workings of coral skeletogenesis and the support it gains from the presence of symbionts. These questions and other recent literature have been reviewed in detail (Buddemeier & Kinzie, 1976). The answers are of substantial importance to the fields of cell biology, geology, paleoclimatology and paleo-oceanography, since they will contribute to both an understanding of ion transport at the cellular level and a knowledge of the environmental-physiological factors reflected in the geochronologies of coral skeletons (Dodge & Vaišnys, 1980).

Several theories of how symbiont photosynthesis affects calcification rate have been proposed. These are roughly divided into (1) those that consider removal of metabolic products from the site of $CaCO_3$ deposition, and (2) those that consider the translocation of photosynthetically fixed carbon to the animal, where it is used either directly as a substrate for skeletal matrix synthesis or indirectly as a general energy source for matrix synthesis or calcium ion movement. None of these possibilities are mutually exclusive, and none have sufficient experimental proof at this time to warrant their adoption (see Buddemeier & Kinzie, 1976). As a general rule, the experimental focus of these studies has followed the calcium ion. Less attention has been given to dissolved inorganic carbon as both a substrate for $CaCO_3$ deposition and a rate limiting factor. Two important questions exist: (1) the identity of the form of dissolved inorganic carbon (CO_2, HCO_3^-, or CO_3^{2-}) supplied to the site of deposition, and (2) the origin of dissolved inorganic carbon utilized in skeletogenesis (i.e., metabolic CO_2 vs. dissolved inorganic carbon in the medium). Comparison of ^{45}Ca and inorganic ^{14}C data suggests that internally produced CO_2 may predominate (Crossland & Barnes, 1974). If this is the case, then symbiont photosynthesis and host calcification are competing, not complementary, processes, and the models used to describe the phenomenon must be reworked accordingly. An increased emphasis on the role of inorganic carbon in symbiotically enhanced calcification is clearly required as an essential step toward balancing the equation. It is also required as an important element in the calculation of primary production by these associations. The importance of the coral skeleton as a sink for inorganic carbon is frequently overlooked and deserves inclusion in the final calculation.

Paragenetic evolution

It is generally recognized that symbiotic events have played an important role in the evolution of the cell, particularly with regard to the origins of cell organelles and the transition between prokaryotes and eukaryotes

(Schwemmler & Schenk, 1980; Frederick, 1981). Less attention has been paid to the impact of a symbiosis on the evolution of multicellular organisms per se, yet it is likely that the pooling of genetic information in the functional symbiotic unit will result in major changes in biological fitness that can be reflected in sudden evolutionary leaps. The mechanism of symbiosis is thus compatible with the punctuational view of evolution (Stanley, 1979) and deserves further consideration in this regard.

The distinction between hermatypic and ahermatypic corals defined by the presence or absence of the symbiont *G. microadriaticum* is obvious. It is generally seen to be the result of an improved nutritional/physiological condition and the photosynthetic enhancement of skeletal deposition. We do not know the extent to which the juxtaposition of algal and coral cells affects gene expression in both organisms. We need to examine more closely the effects that these coexisting genomes have on the development and life histories of both the host and its symbiont, since the developmental and evolutionary consequences are staggering (Taylor, 1981). For this reason, it is important to focus upon the concept of a symbiosis as a functional unit where the sum is generally greater than the parts, and the foundation is intergenomic stimulation and cooperation.

References

Ball, G. (1968). Organisms living on and in protozoa. *In*: T.-T. Chen, ed., *Research in Protozoology*, vol. 3, pp. 566–718. Oxford: Pergamon Press.

Bogorad, L. (1975). Evolution of organelles and eukaryotic genomes. *Science* **188**, 891–8.

Boyle, J. E., & Smith, D. C. (1975). Biochemical interactions between the symbionts of *Convoluta roscoffensis*. *Proc. Royal Soc. Lond.* Series B. **189**, 121–35.

Brandt, K. (1881). Über das Zusammen leben von Thieren und Algen. *Ver. Physiol. Ges. Berlin* **1881–2**, 22–6.

Buddemeier, R. W., & Kinzie, R. A., III. (1976). Coral growth. *Oceanogr. Mar. Biol. Annu. Rev.* **14**, 183–225.

Bunt, J. S. (1969). The CO_2 compensation point, hill activity and photorespiration. *Biochem. Biophys. Res. Comm.* **35**, 748–53.

Bunt, J. S., & Heeb, M. A. (1971). Consumption of O_2 in the light by *Chlorella pyrenoidosa* and *Chlamydomonas reinhardtii*. *Biochim. Biophys. Acta.* **226**, 354–9.

Burris, J. (1977). Photosynthesis, photorespiration, and dark respiration in eight species of algae. *Mar. Biol.* **39**, 371–9.

Chalker, B. E. (1976). Calcium transport during skeletogenesis in hermatypic corals. *Comp. Biochem. Physiol.* **54A**, 455–9.

Chalker, B. E., & Taylor, D. L. (1975). Light enhanced calcification and the role of oxidative phosphorylation in calcification of the coral *Acropora cervicornis*. *Proc. Royal Soc. Lond.* Series B. **190**, 323–31.

Crossland, C. J., & Barnes, D. J. (1974). The role of metabolic nitrogen in coral calcification. *Mar. Biol.* **28**, 325–32.

Deane, E. M., & O'Brien, R. W. (1981a). Uptake of sulfate, taurine, cysteine and methionine by symbiotic and free-living dinoflagellates. *Arch. Microbiol.* **128**, 311–19.
– (1981b). Uptake of phosphate by symbiotic and free-living dinoflagellates. *Arch. Microbiol.* **128**, 307–10.
D'Elia, C. F. (1977). The uptake and release of dissolved phosphorus by reef corals. *Limnol. Oceanogr.* **22**, 301–15.
D'Elia, C. F., & Webb, K. L. (1977). The dissolved nitrogen flux of reef corals. *Proc. 3d. Int. Coral Reef Symp.* **1**, 325–30.
DiSalvo, L. H. (1971). Ingestion and assimilation of bacteria by two scleractinian coral species. *In*: H. M. Lenhoff, L. Muscatine, & L. V. Davies, eds., *Experimental Coelenterate Biology*, pp. 129–39. Honolulu: University of Hawaii Press.
Dodge, R. E., & Vaišnys, J. R. (1980). Skeletal growth chronologies of recent and fossil corals. *In*: D. C. Rhoads & R. A. Lutz, eds., *Skeletal Growth of Aquatic Organisms*, pp. 493–518. New York: Plenum.
Ducklow, H. W., & Mitchell, R. (1979a). Bacterial populations and adaptations in the mucus layers of living corals. *Limnol. Oceanogr.* **24**, 715–25.
– (1979b). Composition of mucus released by coral reef coelenterates. *Limnol. Oceanogr.* **24**, 706–14.
Fredrick, J. F., ed. (1981). Origins and evolution of eukaryotic intracellular organelles. *Ann. N.Y. Acad. Sci.* **361**, 1–510.
Gibbs, S. P. (1981). The chloroplasts of some algal groups may have evolved from endosymbiotic eukaryotic algae. *Ann. N.Y. Acad. Sci.* **361**, 193–208.
Glynn, P. W. (1973). Ecology of a Caribbean coral reef. The *Porites* reef-flat biotype. II. Plankton community with evidence for depletion. *Mar. Biol.* **22**, 1–21.
Gooday, G. W., & Doonan, S. A. (1980). The ecology of algal-invertebrate symbioses. *Contemporary Microbial Ecology. Proc. 2nd Int. Symp. on Microbial Ecology*, pp. 377–390. London: Academic Press.
Goreau, T. F., & Goreau, N. I. (1959). The physiology of skeleton formation in corals. II. Calcium deposition by hermatypic corals under various conditions on the reef. *Biol. Bull. Mar. Biol. Lab., Woods Hole.* **117**, 239–50.
Hollande, A., & Carré, D. (1975). Les xanthelles des radiolaires sphaerocollides, des acanthaires et de *Velella velella*: Infrastructure – cytochime – taxonomie. *Protistologica* **10**, 576–601.
Jeffrey, S. W., & Vesk, M. (1976). Further evidence for a membrane-bound endosymbiont within the dinoflagellate *Peridinium foliaceum*. *J. Phycol.* **12**, 450–5.
Johannes, R. E., Coles, S. L., & Kuenzel, N. T. (1970). The role of zooplankton in the nutrition of some scleractinian corals. *Limnol. Oceanogr.* **15**, 579–86.
Kawaguti, S., & Sakumoto, D. (1948). The effect of light on the calcium deposition of corals. *Bull. Oceanogr. Inst. Taiwan* **4**, 65–70.
Keeble, F. W. (1910). *Plant Animals*. Cambridge: Cambridge University Press.
Kollar, E. J., & Fisher, C. (1980). Tooth induction in chick epithelium: Expression of quiescent genes for enamel synthesis. *Science* **207**, 993–5.
Lasker, H. R. (1981). Phenotypic variation in the coral *Montastrea cavernosa* and its effects on colony energetics. *Biol. Bull. Mar. Biol. Lab., Woods Hole* **160**, 292–302.
Lewis, D. H., & Smith, D. C. (1971). The autotrophic nutrition of symbiotic marine coelenterates with special reference to hermatypic corals. I. Movement of photosynthetic products between the symbionts. *Proc. Royal Soc. Lond.* Series B. **178**, 111–29.

Lewis, J. B. (1976). Experimental tests of suspension feeding in Atlantic reef corals. *Mar. Biol.* **36**, 147–50.
Lewis, J. B., & Price, W. S. (1975). Feeding mechanisms and feeding strategies of Atlantic reef corals. *J. Zool., Lond.* **176**, 527–44.
Loeblich, A. R., III, & Sherley, J. L. (1979). Observations on the theca of the motile phase of free-living and symbiotic isolates of *Zooxanthella microadriatica* (Freudenthal) comb. nov. *J. Mar. Biol. Ass., U.K.* **59**, 195–206.
McCloskey, L. R., Wethey, D. S., & Porter, J. W. (1978). The measurements and interpretation of photosynthesis and respiration in reef corals. *Monogr. Oceanogr. Methodol.* **5**, 379–96.
Meyers, J. (1980). On the algae: Thoughts about physiology and measurements of efficiency. *In*: P. Falkowski, ed., *Primary Productivity in the Sea*, pp. 1–16. New York: Plenum Press.
Muscatine, L. (1967). Glycerol excretion by symbiotic algae from corals and *Tridacna* and its control by the host. *Science* **156**, 516–19.
– (1973). Nutrition of corals. *In*: O. A. Jones & R. Endean, eds., *Biology and Geology of Coral Reefs*, vol. 2, pp. 77–115. New York: Academic Press.
– (1980). Productivity of zooxanthellae. *In*: P. Falkowski, ed., *Primary Productivity in the Sea*, pp. 381–402. New York: Plenum.
Muscatine, L., & Cernichiari, E. (1969). Assimilation of photosynthetic products of zooxanthellae by a reef coral. *Biol. Bull. Mar. Biol. Lab., Woods Hole,* **137**, 506–23.
Muscatine, L., & Porter, J. W. (1977). Reef corals: Mutualistic symbioses adapted to nutrient-poor environments. *BioScience* **27**, 454–60.
Muscatine, L., McCloskey, L. R., & Marian, R. E. (1981). Estimating the daily contribution of carbon from zooxanthellae to coral animal respiration. *Limnol. Oceanogr.* **26**, 601–11.
Pearse, V. B., & Muscatine, L. (1971). Role of symbiotic algae (zooxanthellae) in coral calcification. *Biol. Bull. Mar. Biol. Lab., Woods Hole,* **141**, 350–63.
Roushdy, H. M., & Hansen, V. K. (1961). Filtration of phytoplankton by the octocoral *Alcyonium digitatum* L. *Nature Lond.* **190**, 649–50.
Rublee, P. A., Lasker, H. R., Gottfried, M., & Roman, M. R. (1980). Production and bacterial colonization of mucus from the soft coral *Briarium abestinum. Bull. Mar. Sci.* **30**, 888–93.
Schmitz, K., & Kremer, B. P. (1977). Carbon fixation and analysis of assimilates in a coral dinoflagellate symbiosis. *Mar. Biol.* **42**, 305–13.
Schoenberg, D. A., & Trench, R. K. (1980a). Genetic variation in *Symbiodinium* (= *Gymnodinium*) *microadriaticum* Freudenthal, and specificity in its symbiosis with marine invertebrates. I. Isoenzyme and soluble protein patterns of axenic cultures of *Symbiodinium microadriaticum. Proc. Royal Soc. Lond.* Series B. **207**, 405–27.
– (1980b). Genetic variation in *Symbiodinium* (= *Gymnodinium*) *microadriaticum* Freudenthal, and specificity in its symbiosis with marine invertebrates. II. Morphological variation in *Symbiodinium microadriaticum. Proc. Royal Soc. Lond.* Series B. **207**, 429–44.
– (1980c). Genetic variation in *Symbiodinium* (= *Gymnodinium*) *microadriaticum* Freudenthal, and specificity in its symbiosis with marine invertebrates. III. Specificity and infectivity of *Symbiodinium microadriaticum. Proc. Royal Soc. Lond.* Series B. **207**, 445–60.
Schwemmler, W., & Schenk, H. E. A., eds. (1980). *Endocytobiology, Endosymbiosis and Cell Biology*. Berlin: Walter de Gruyter & Co.
Smith, D. C. (1979). From extracellular to intracellular: the establishment of a symbiosis. *Proc. Royal Soc. Lond.* Series B. **204**, 115–30.

Smith, D. C., Muscatine, L., & Lewis, D. H. (1969). Carbohydrate movement from autotrophs to heterotrophs in parasitic and mutualistic symbiosis. *Biol. Rev.* **44**, 17–90.

Smith, S. V. (1978). Coral-reef area and the contributions of reefs to processes and resources of the world's oceans. *Nature* **273**, 225–6.

Sorokin, Y. I. (1973). On the feeding of some scleractinian corals with bacteria and dissolved organic matter. *Limnol. Oceanogr.* **18**, 380–5.

Sournia, A. (1977). Analyse et bilan de la production primaire dans les récifs coralliens. *Ann. l'Inst. Océanographique, Paris* **53**, 47–74.

Stanley, S. M. (1979). *Macroevolution: Pattern and Process.* San Francisco: Freeman.

Stephens, G. C. (1962). Uptake of organic material by aquatic invertebrates. I. Uptake of glucose by the solitary coral *Fungia scutaria. Biol. Bull. Mar. Biol. Lab., Woods Hole* **123**, 648–59.

Summons, R. E., & Osmond, C. B. (1981). Nitrogen assimilation in the symbiotic marine alga *Gymnodinium microadriaticum*: Direct analysis of ^{15}N incorporation by GC-MS methods. *Phytochemistry* **20**, 575–8.

Taylor, D. L. (1971). Ultrastructure of the "zooxanthellae" *Endodinium chattoni* Hovasse *in situ. J. Mar. Biol. Ass., U.K.* **51**, 227–34.

– (1973a). Cellular interactions of algal-invertebrate symbiosis. *Adv. Mar. Biol.* **11**, 1–56.

– (1973b). Algal symbionts of invertebrates. *Ann. Rev. Microbiol.* **27**, 171–87.

– (1974a). Symbiotic marine algae: Taxonomy and biological fitness. *In*: W. B. Vernberg & F. J. Vernberg, eds., *Symbiosis in the Sea*, pp. 245–62. Columbia: University of South Carolina Press.

– (1974b). Nutrition of algal-invertebrate symbiosis. I. Utilization of soluble organic nutrients by symbiont-free hosts. *Proc. Royal Soc. Lond.* Series B. **186**, 357–68.

– (1975). Symbiotic dinoflagellates. *Symp. Soc. Exper. Biol.* **29**, 267–78.

– (1978a). Artificially induced symbiosis between marine flagellates and vertebrate tissues in culture. *J. Protozool.* **25**, 77–81.

– (1978b). Nutrition of algal-invertebrate symbiosis. II. Effects of exogenous nitrogen sources on growth, photosynthesis and the rate of excretion by algal symbionts *in vivo* and *in vitro. Proc. Royal Soc. Lond.* Series B. **201**, 401–12.

– (1980). Nutrient competition as a basis for symbiont selection in algal-invertebrate symbiosis. *In*: W. S. Schwemmler & H. Schenk, eds., *Endocytobiology, Endosymbiosis and Cell Biology*, vol. 1, pp. 265–71. Berlin: Walter de Gruyter & Co.

– (1981). The evolutionary impact of intracellular symbiosis. *Ber. Deutsch. Bot. Ges.* **94**, 583–90.

– (in press). Nutrition of algal-invertebrate symbiosis. *C.R.C. Handbook on Nutrition and Food.* C.R.C. Press.

Tomas, R. N., & Cox, E. R. (1973). Observations on the symbiosis of *Peridinium balticum* and its intracellular alga. I. Ultrastructure. *J. Phycol.* **9**, 304–23.

Trench, R. K. (1979). The cell biology of plant-animal symbiosis. *Ann. Rev. Plant Physiol.* **30**, 485–531.

Vandermeulen, J. H., Davis, N., & Muscatine, L. (1972). The effect of inhibitors of photosynthesis on zooxanthellae in corals and other marine invertebrates. *Mar. Biol.* **16**, 185–91.

Yonge, C. M. (1930). Studies on the physiology of corals. I. Feeding mechanisms and food. *Rep. Great Barrier Reef Exped.* **1**, 14–57.

– (1936). Mode of life, feeding, digestion and symbiosis with zooxanthellae in the Tridacnidae. *Rep. Great Barrier Reef Exped.* 1, 283–321.
– (1958). Ecology and physiology of reef building corals. *In*: A. A. Buzzati-Traverso, ed., *Perspectives in Marine Biology*, pp. 117–35. Berkeley: University of California Press.

3

Symbiosis in foraminifera

JOHN J. LEE
Department of Invertebrates
American Museum of Natural History
Central Park West at 79th St.
New York, NY 10024

MARIE E. McENERY
Department of Biology
City College of CUNY
Convent Ave. at 138th St.
New York, NY 10031

Protozoologists and micropaleontologists have long known that there are unusually large or even giant (some have volumes that exceed 500 mm^3) foraminifera in shallow benthic tropical and semitropical waters. The modern larger species are so diverse that they are placed into ten distinct families (Ross 1974). An additional 12 families have been recognized in the fossil record where specimens of one of these protozoan giants, *Nummulities gizehensis*, reached diameters of more than 120 mm in the Eocene (Newell 1949). Though some of the earliest students of living foraminifera (e.g., Butschli 1886, Cushman 1922, Rhumbler 1909, Winter 1907) commented on the presence of algae within some foraminifera, it has been generally appreciated only recently that all of the modern larger foraminifera are the hosts for endosymbiotic algae (Lee 1974, 1980, Lee et al. 1980, Ross 1974). Present evidence also suggests that adaptation for symbiosis has been a key driving force in the evolution of larger foraminifera in shallow tropical waters (Lee 1974, Lee et al. 1979, Ross 1974).

Larger foraminifera are quite abundant in the euphotic shallow benthic zones of modern tropical and semitropical seas (Brasier 1975a, b, Hansen & Buchardt 1977, Lutze 1980, Ross 1974, Smith & Wiebe 1977, Sournia 1976, 1977) where they play important roles in biogeochemical mineral cycling. Evidence from Takapoto Atoll, Tuamotu Islands in French Polynesia obtained by Sournia (1976) suggests that the foraminiferal sands in the lagoon were 50 times more productive than the plankton, fixing approximately 430–1,330 mg C m^{-2} day^{-1}. The same worker (Sournia 1977) found a similar situation in the Gulf of Elat on the Red Sea. There

The authors wish to acknowledge the generous support of the National Science Foundation (OCE 78 25946) for their work on symbiosis in foraminifera. The optics used in our studies were gifts from the H. B. Cantor Foundation.

This paper is an abridgment of a lecture given at the dedication of the H. B. Cantor Microbiology Resource Room at the City College.

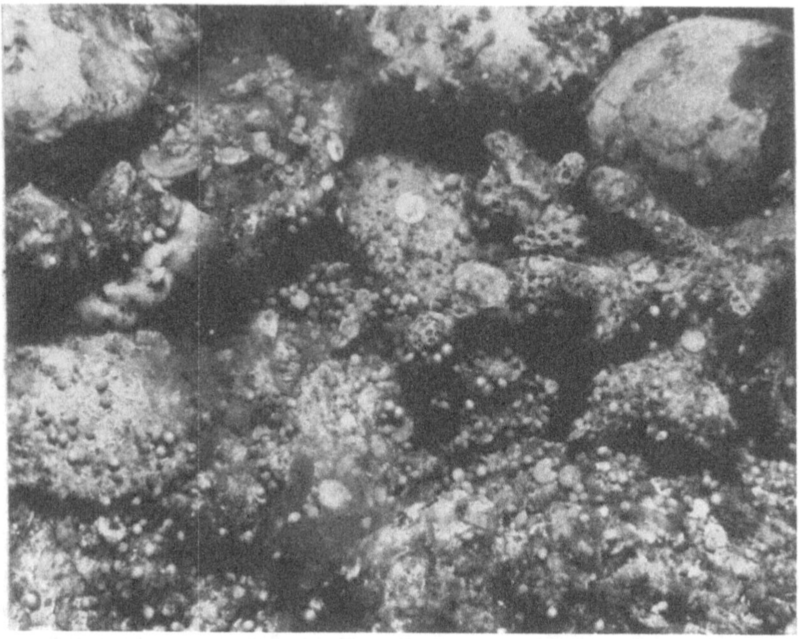

Fig. 1. Larger foraminifera on the surface of benthic granite and coral rubble. Photo taken at 25 m in the Gulf of Elat, Red Sea, very near the Heintz Steinitz Marine Biology Laboratory of the Hebrew University of Jerusalem. The seed-shaped foraminifera are *Amphistegina* spp.; the coin-shaped foraminifera are *A. hemprichii*. Enlarged in printing 1.7×.

he found an average 3–4 μg chlorophyll *a*/g dry weight benthic sediment, more than ten times the level he found in the phytoplankton during the month he sampled (May 1978). In her study of production in a tide pool at Makapuu Point, Hawaii, Muller (1974) estimated that about 1×10^8 new tests (5×10^2 g $CaCO_3$)m^{-2} of *Amphistegina* were added to the sediments each year.

Larger foraminifera, in common with foraminifera in many habitats (e.g., Buzas 1968, 1969, 1970, Lee et al. 1969, Lynts 1966, Matera & Lee 1972), are very heterogeneous in their distribution. This phenomenon has seldom been documented or systematically investigated in most of the tropical and subtropical habitats where larger foraminifera abound. Ross (1972) noted that the larger foraminifer, *Marginopora vertebralis*, was quite patchy in its distribution on the Great Barrier Reef. He found an average of 20–30 animals 12 mm or larger per square meter; however, densities reached 100 animals m^{-2} in patches near the Crown of Thorns Research Station. In the Gulf of Elat local patches of the larger foraminifera *Amphisorus hemprichii* and *Amphistegina* spp. reach even higher densities. In some patches they reach densities greater than $10^5 - 10^6$

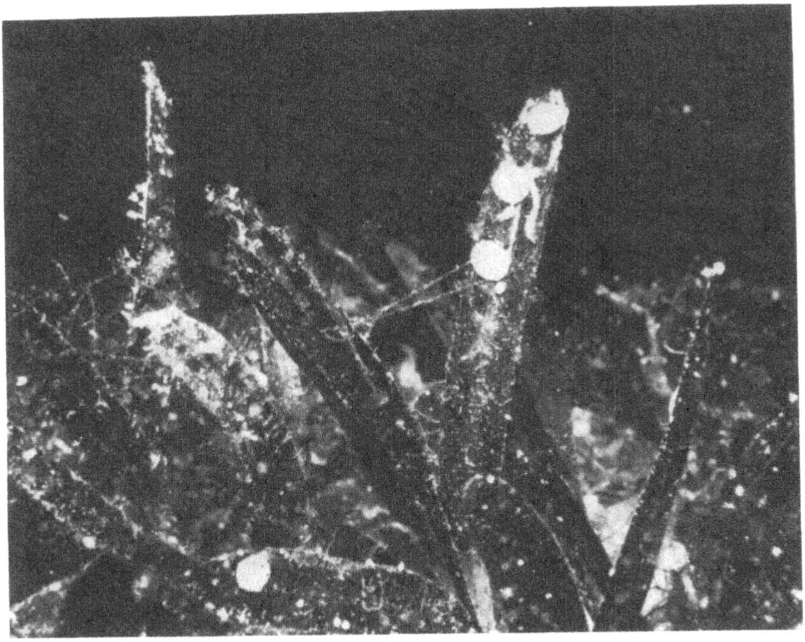

Fig. 2. *A. hemprichii* and *Amphistegina* spp. on the surface of *Halophila*, a benthic sea grass. Photo taken at 20 m near the same site as Fig. 1. Enlarged in printing 1.7×.

m^{-2} (Figs. 1 and 2) while at the same time it is difficult to locate them by scuba diving only a few meters away. We have noted the same distributional pattern in the Florida Keys (Lee & Zucker 1969). There *Archaias angulatus* and *Sorites marginalis* reach similar high densities as epibionts on patches of *Thalassia testudinum* growing in shallow (1–4 m) embayments (e.g., Biscayne Bay, Key Largo Sound).

Though we know more about endosymbiosis in larger foraminifera, the phenomenon is not restricted to this group of foraminifera. Algal symbionts are found in some planktonic (globigerinid) species (Alldredge & Jones 1973, Anderson & Bé 1976a, Bé et al. 1977, Lee et al. 1965, Zucker 1973). Recent ultrastructural, pigment, and radiotracer studies of three brackish water species, *Elphidium williamsoni, E. excavatum*, and *Nonion germanicum*, from Danish water stations suggest that chloroplasts from ingested diatoms or chrysophytes may not be digested at the same rate as the rest of the algal cell. In two of the three foram species, *E. williamsoni* and *N. germanicum*, the algal chloroplasts persist and apparently function photosynthetically (Lopez 1979).

Algae involved in foraminiferal endosymbioses

By comparison to other groups (e.g., corals) the modern larger foraminifera are amazingly catholic with respect to the types and species of

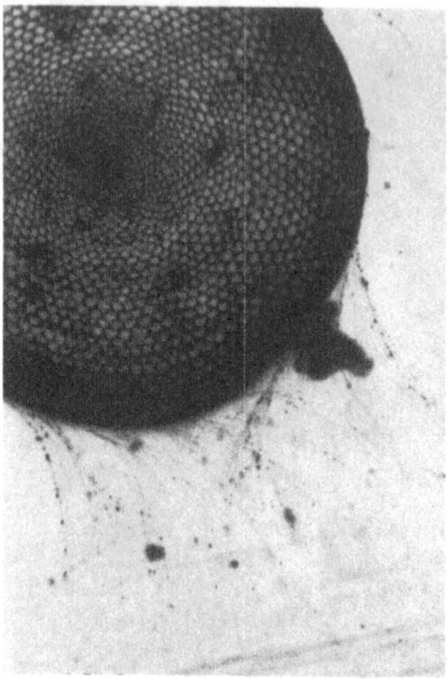

Fig. 3. Photomicrograph of *A. hemprichii* from a laboratory culture showing pseudopodial food web. 10×.

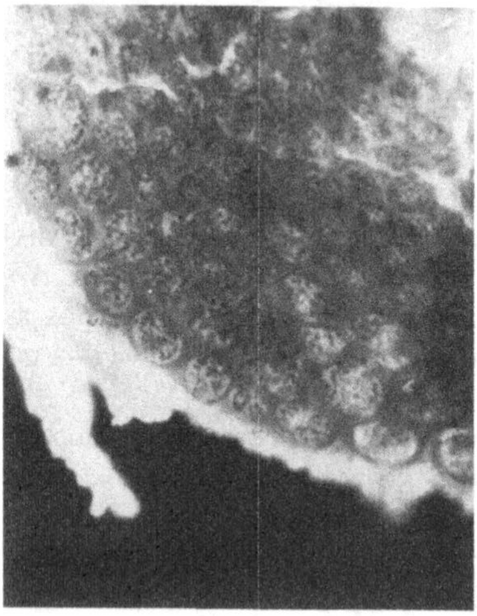

Fig. 4. Transmission light microscopic (TLM) photograph showing a portion of *S. marginalis* with dinoflagellate endosymbionts within chamberlets. Shell has been partially broken (upper half of photo) in order to flatten specimen for photography. Photograph taken through the thin upper test wall. 35×.

algae that are collectively husbanded by the group. Different species have chlorophytes, dinoflagellates, diatoms, or rhodophytes as endosymbionts. The possibility of a cryptophyte endosymbiont (Winter 1907) requires confirmation.

The common dinoflagellate endosymbiont of corals, other coelenterates, and molluscs, *Symbiodinium microadriaticum*, also seems to be the endosymbiont in a number of species of Soritidae including *A. hemprichii* (Leutenegger 1977a, b) (Fig. 3), S. *marginalis* (Müller-Merz & Lee 1976)

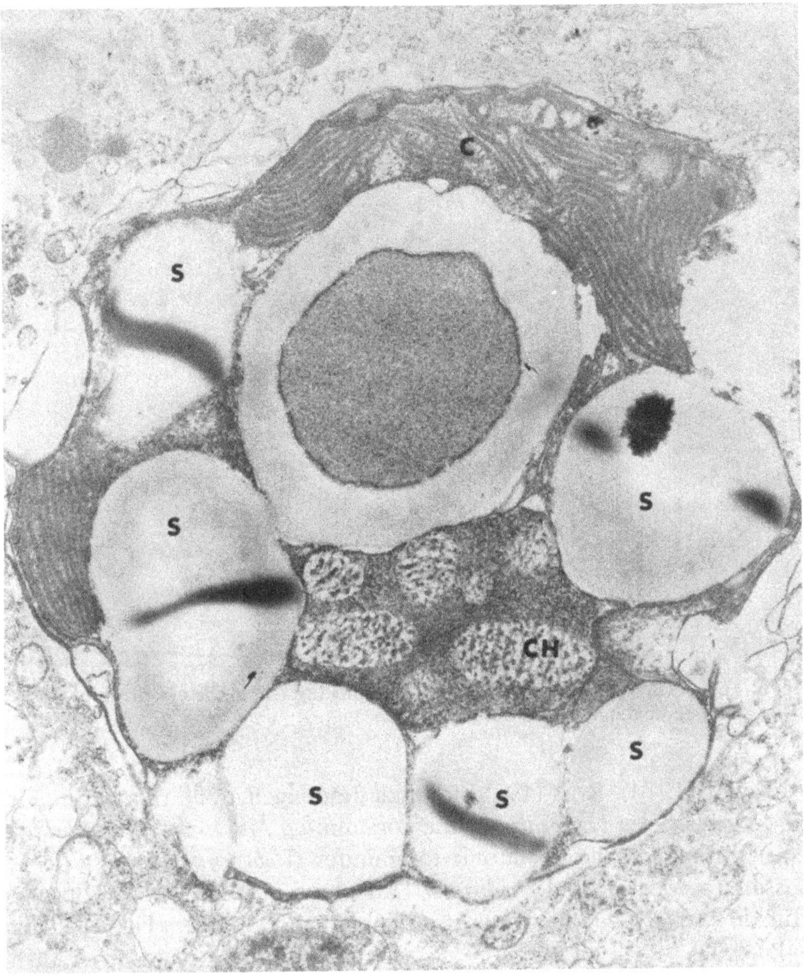

Fig. 5. Transmission electron microscope (TEM) photograph showing *Zooxanthella microadriatica* within the cytoplasm of *S. marginalis*. Photo by Dr. Edith Müller-Merz, Dept. of Biology, City College. S, starch; C, chloroplast; CH, chromosome. 20,000×.

(Fig. 4), *S. orbiculus* (Leutenegger 1977a, b), and *S. orbitolites* (Leutenegger 1977b, c). Cells of *S. microadriaticum* in hosts that are freshly fixed shortly after collection are filled with starch grains (Fig. 5). Recent morphological, biochemical, and infectivity studies on *S. microadriaticum* from coelenterate and molluscan hosts have raised new issues (Loeblich & Sherley 1979, Schoenberg 1980a, b, Schoenberg & Trench 1980a–c) of strain affinities (see Chapter 2, this volume). However, there are no data on these points for the foraminiferal dinoflagellate endosymbionts. Species of *Amphidinium* were found in *Peneroplis* sp. and *Orbitolites* sp.

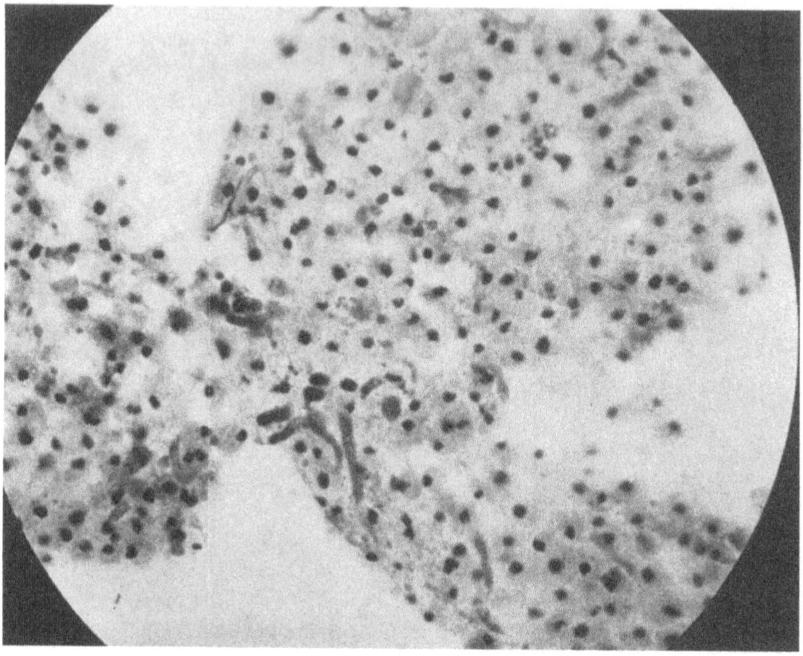

Fig. 6. TLM section showing dinoflagellate zooxanthellae within the cytoplasm of *Globigerina ruber*. Photography by the authors. 400×.

by Taylor (1974). Ross (1974) found a dinoflagellate he thought might be *Gymnodinium vertebralis* in the foraminifer *Marginopora vertebralis*. The symbiont in the planktonic foraminifer *Globigerinoides sacculifer* is possibly a species of *Aureodinium* (Anderson and Bé 1976b, Bé and Hutson 1977). The identity of the symbionts in *G. ruber* (Fig. 6) is still unknown.

Living chlorophytes have been isolated from seven species of larger foraminifera. Two of the foraminiferal species, *A. angulatus* and *Cyclorbiculina compressa*, from the very shallow waters (1–3 m) of Key Largo Sound and Biscayne Bay, Florida, are always intensely green due to their respective chlorophyte endosymbionts, *Chlamydomonas hedleyi* and *C. provasolii* (Lee & Zucker 1969, Lee et al. 1974, 1979). These two *Chlamydomonas* species are easily distinguished from each other in the transmission electron microscope (TEM) in situ and in vitro if one can view sections that include pyrenoids (Figs. 7, 8). The pyrenoid of one species, *C. provasolii*, is typically penetrated by many tubules each with a single thylakoid stack in the center. The pyrenoid also has a central fibrous knot and is surrounded by an extensive starch sheath with many small starch grains (Fig. 8). The pyrenoid of *C. hedleyi*, on the other hand, is pierced by a single tubule and is surrounded by a starch sheath typically

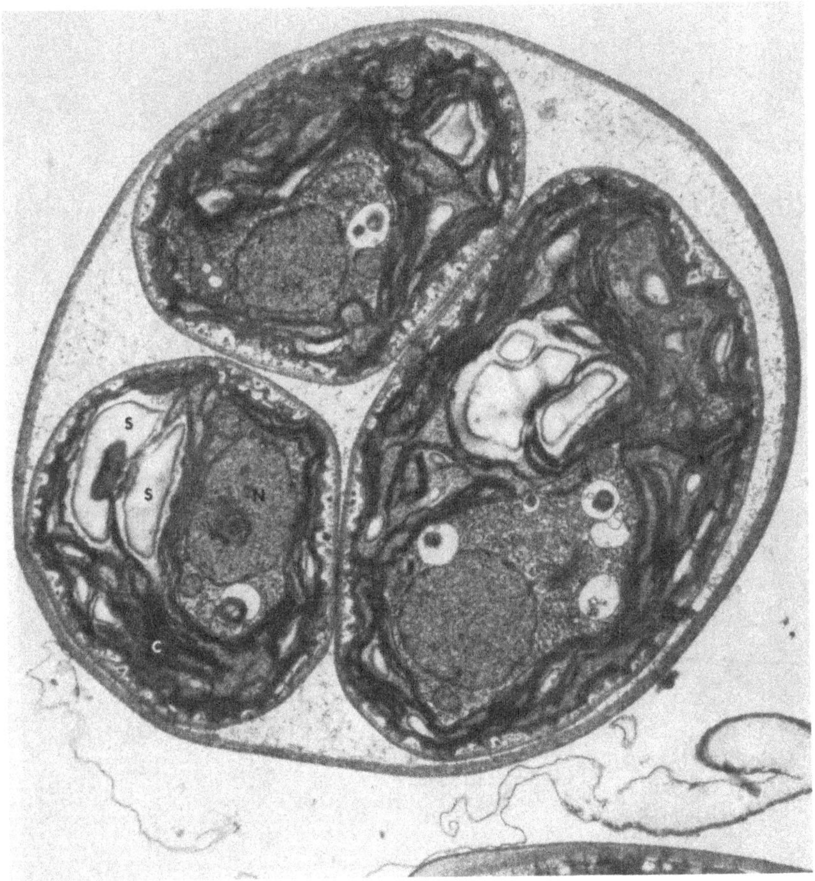

Fig. 7. TEM photograph of *C. hedleyi* from culture. Photograph by Dr. Frederick Schuster, Department of Biology, Brooklyn College. N, nucleus; S, starch within pyrenoid; C, chloroplast. 9,500 ×.

composed of two large starch grains (Fig. 7). *C. provasolii* has also been identified as part of the internal flora of *S. marginalis* in TEM sections where its numbers are low in comparison to *S. microadriaticum* (Müller-Merz & Lee 1976).

Pigment analyses of two species of larger foraminifera, *A. hemprichii* and *Amphistegina lobifera*, from shallow waters in the Gulf of Elat disclosed small amounts of chlorophyll *b* and suggested that these animals might harbor some chlorophytes (Lee et al. 1980a). The algae have been isolated subsequently from a few specimens each of *A. hemprichii, A. lobifera, A. lessonii, A. papillosa*, and *Heterostegina depressa*. Morphological, fine structural and culture observations have shown that the algae are a species of *Chlorella* (Fig. 9) (Lee et al. 1980d). Specimens of *H. depressa*,

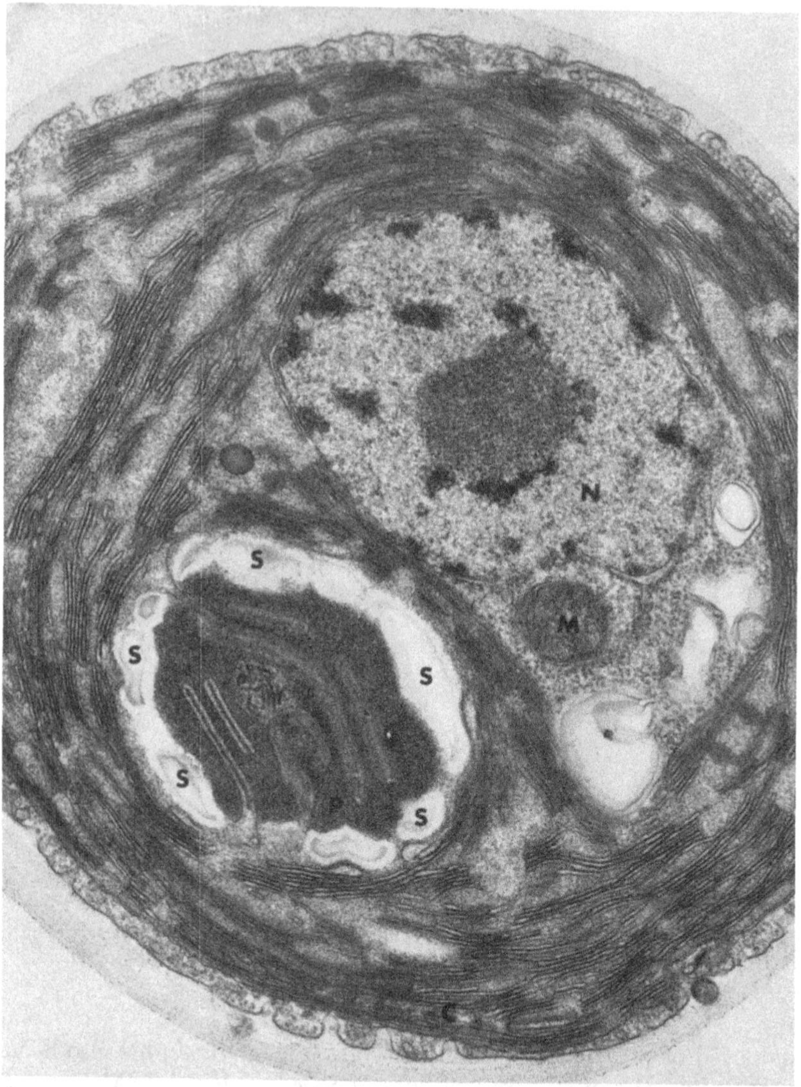

Fig. 8. TEM photograph showing *C. provasolii* from culture. Photograph by Dr. Frederick Schuster, Department of Biology, Brooklyn College. C, chloroplast; N, nucleus; M, mitochondria; S, starch within pyrenoid. 23,500×.

captured near Maui Island, Hawaii, harbored a different species of *Chlorella*. The two *Chlorella* species differ from each other in culture characteristics, fine structure of the cell wall, chloroplast and pyrenoid structure, and autospore production (J. J. Lee et al. 1980, 1982) and differ morphologically from the "*Chlorella*-like" algae recently described from the marine coelenterate *Anthopleura xanthogrammica* (O'Brien 1978).

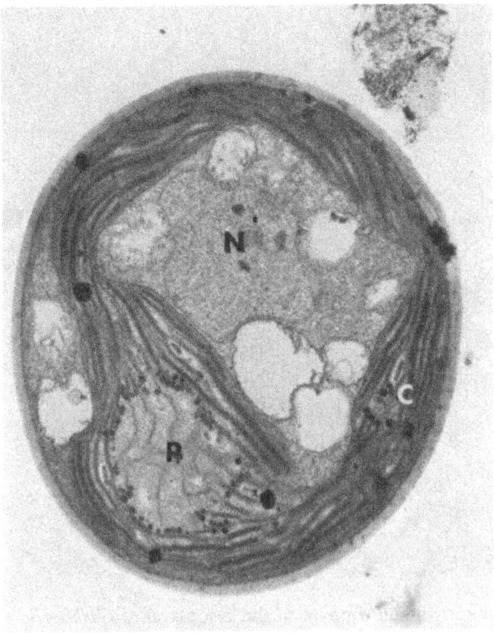

Fig. 9. TEM photograph of *Chlamydomonas* sp. isolated in culture from *A. lessonii*. Photograph by Jason Reidy, Department of Biology, City College. C, chloroplast; P, pyrenoid; N, nucleus. 10,000×.

Biochemical and physiological tests using the methodology of Kessler and his co-workers (Kessler 1975, 1976, 1978, Kessler & Czygan 1970, Kessler & Zweier 1971) have been performed on both species of *Chlorella*. Using the criteria developed by Kessler (1976, 1978) the alga isolated from the Hawaiian strain of *H. depressa* is clearly a strain of *C. saccharophila* (J. J. Lee et al. 1982). The other *Chlorella* sp. belongs to a different taxon. Because it failed to grow in both the nitrate medium of Kessler and Czygan (1970) and the thiamine-supplemented ammonium medium of Kessler and Zweier (1971) it could not be compared to other species biochemically and physiologically characterized by Kessler (1976, 1978, J. J. Lee et al. 1982). Since the *Chlorella* spp. were not observed in the thin sections of the foregoing species of foraminifera studied in the TEM (Leutenegger 1977a–c), one must conclude that they are of only minor importance, if any, to their hosts. Although it is tempting to suggest that the rare chlorophytes may be undigested food in vacuoles or culture artifacts, recent work in progress may indicate otherwise. In attempts to isolate in culture the dinoflagellate symbionts from several globigerinids, we also isolated a chlorophyte. Looking back through TEM micrographs, Anderson (unpublished) was able to identify the intact chlorophytes, which were overshadowed by the presence of so many dinoflagellate symbionts (Fig. 10).

Diatoms are the dominant algal endosymbionts among the larger for-

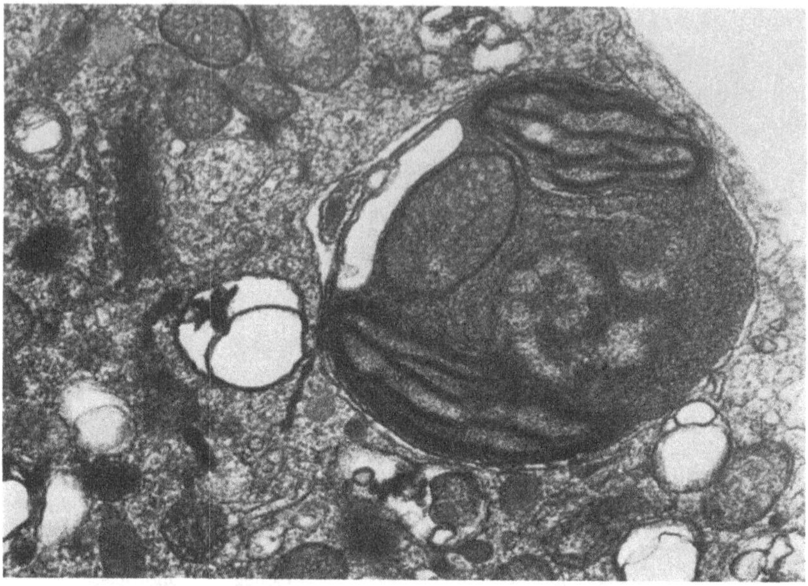

Fig. 10. TEM section showing an unknown chlorophyte in the cytoplasm of *Globigerinoides glutinata*. This organism has been recently isolated in culture. Photograph by Dr. O. Roger Anderson, Lamont-Doherty Geological Observatory, Columbia University. 1,100×.

aminifera belonging to the families Amphisteginidae and Nummulitidae that have been examined thus far. Fine structural studies of a number of species in the TEM suggested that the endosymbionts were most probably frustuleless diatoms (Berthold 1978, Leutenegger 1977a, b, Schmaljohann and Röttger 1976, 1978), but this was not confirmed until they were isolated in culture. Fortunately the endosymbionts isolated in culture formed frustules; otherwise they could not have been identified since it is the characteristics of the frustule that are used in diatom taxonomy. The diatoms isolated from different species of larger foraminifera belong to a number of pennate genera: *Fragilaria, Navicula, Nitzschia*, and *Amphora* (reviewed in Lee 1980). *Fragilaria shiloi* (Fig. 11), one of the smallest pennate diatoms known, was the most abundant endosymbiont in three species of *Amphistegina* (*A. lobifera, A. lessonii*, and *A. papillosa*) examined from the Gulf of Elat in the Red Sea (Lee et al. 1980b). It was also isolated from *A. lobifera* collected in Hawaiian waters (Lee et al. 1980c). However, the specimens of *A. lessonii* from Hawaii harbored a different endosymbiotic diatom, *Nitzschia laevis* (Fig. 12). In addition to their primary symbionts, *Amphistegina* spp. specimens from the Red Sea occasionally ($\lessapprox$20% of the animals) were the hosts for second algal species. These less numerous algae were *Chlorella* sp. or three

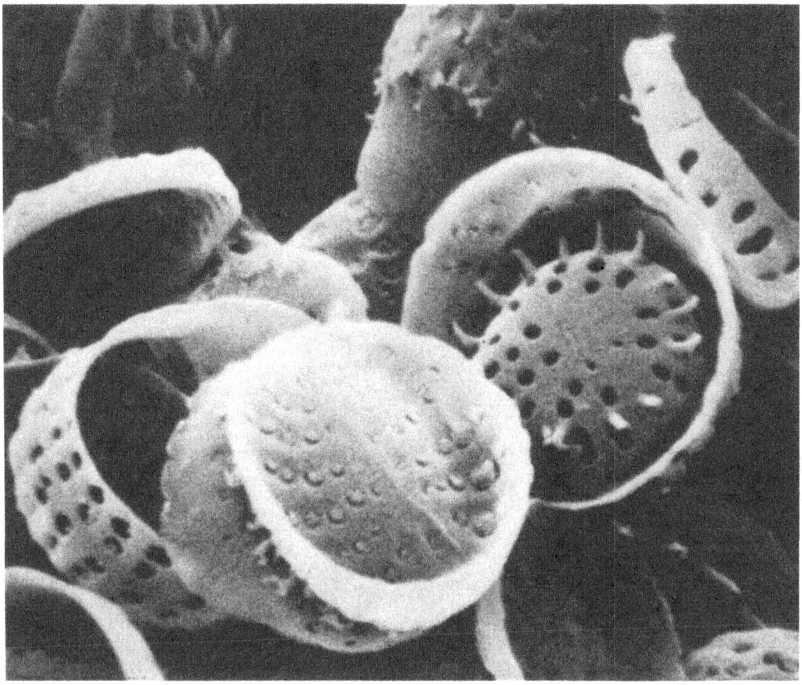

Fig. 11. SEM photograph of *F. shiloi* from culture. This symbiotic diatom was isolated from *Amphistegina* spp. from the Gulf of Elat and from *A. lobifera* from Hawaii. 10,000×.

other diatom species: *Nitzschia panduriformis* var. *continua, Nitzschia frustulum* var. *subsalina*, or *Navicula reissi*. Although the fine structural studies of three species of modern nummulitids, *Operculina ammonoides, Heterocyclina tuberculata*, and *H. depressa*, suggest that these animals also are hosts for endosymbiotic diatoms (Berthold 1978, Leutenegger 1977a, b, Schmaljohann & Röttger 1976, 1978), only the latter species has been examined by gnotobiotic techniques (Lee et al. 1980 a–c). *N. panduriformis* var. *continua* (Fig. 13) was isolated from specimens of *H. depressa* collected in the Red Sea whereas *N. valdestriata* was isolated from Hawaiian specimens.

The published evidence to date and work in progress (e.g. Lee & Reimer, in press) in our laboratory have not yet yielded a definitive picture of the range of diatom species that can be husbanded by particular species of larger foraminifera. It is clear that populations of larger foraminifera from a single collection at one station all husband the same endosymbiotic diatom species. Specimens of the same foram species collected from other stations or the same station, months or years later, may or may not harbor the same endosymbiotic diatom species. Indi-

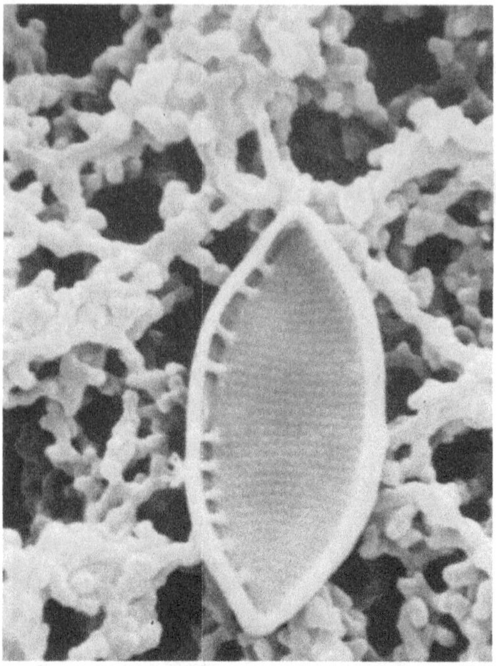

Fig. 12. SEM photograph of *N. laevis* from culture. This symbiotic diatom was isolated from *A. lessonii* from Hawaii. 5,000×.

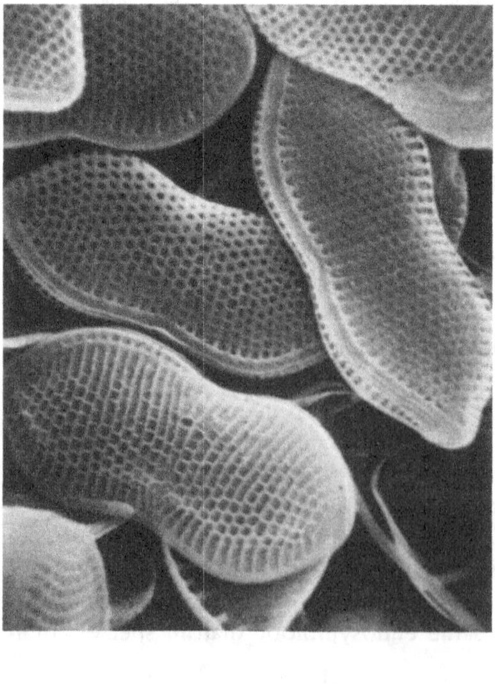

Fig. 13. SEM photograph of *N. panduriformis* var. *continua* from culture. This symbiotic diatom was isolated from *H. depressa* from the Red Sea. 3,600×.

vidual animals usually are the host for only one diatom species at a time. A third or less of the animals may harbor a second, very much less abundant diatom species (Lee 1980). Whether the differences in the diatom species harbored by a single foraminiferal species are a matter of selective or stochastic recruitment by zygotes in the sexual phase of the life cycle or physiological responses to abiotic factors (e.g., light, depth) is not yet known. These and related questions are under active investigation in our laboratory.

A unicellular rhodophyte, possibly *Porphyridium* sp. has been observed by light and transmission electron microscopy in specimens of *Peneroplis planatus* and *Spirolina* sp. (Lee unpublished, Leutenegger 1977b, Röttger et al. 1980). The alga has just been axenically cultivated (work in progress in our lab), but has not yet been totally characterized.

Morphological and physiological characteristics of the algae

Symbiotic algae, within the cytoplasm of their host foraminifera, are easily distinguished in transmission electron micrographs from algae in food vacuoles. Although both are bounded by both host membranes and their own cell membranes, the symbiotic algae lose their characteristic cell envelopes (e.g., frustules, cell walls, thecae). This is quite frustrating if one is seeking answers to taxonomic questions, because the best one can expect to obtain from fine structural studies or pigment extractions is a general categorization of algal type (e.g., Berthold 1978, Leutenegger 1977a, b, Schmaljohann & Röttger 1976, 1978). Fortunately algae isolated from their hosts regain their abilities to form their envelopes and other diagnostic features, which are needed to fully characterize and identify them (Lee et al. 1980a–c).

As a group, the algae symbiotic in foraminifera have phosphorus and nitrogen requirements for optimal growth that greatly exceed the levels available in the free living environment where their hosts are collected (Lee et al. 1979, 1980d). *N. panduriformis* var. *continua* from *H. depressa* collected in Elat grew best in a medium containing a concentration of 2 μM $NaNO_3$ as a nitrogen source whereas the concentration of NO_2^{1-} and NO_3^{1-} in the Gulf of Elat rarely exceeds 1 μg-at/l at the depth where the animals were collected (Levanson-Spanier et al. 1979). *N. laevis, N. frustulum* var. *subsalina, N. valdestriata*, and *F. shiloi* needed even higher concentrations of $NaNO_3$ (2 m*M*) for maximum growth rates (Lee et al. 1980d).

With respect to phosphorus requirements, the optimum growth rates of these algae were obtained between 1 and 100 μ*M* Na glycerol phosphate whereas the levels of phosphate in the Gulf rarely exceed 0.3 μg-at/l at the same depth. The results of these experiments suggest that either the foraminiferal-algal symbiotic systems tightly recycle nitrogen and phosphorus or they are constantly nutrient limited.

Most of the axenic cultures of the endosymbiotic algae isolated from larger foraminifera require or are stimulated by vitamin B_{12}, biotin, and/or thiamine (Lee et al. 1979, 1980a, b, c). The growth of *C. provasolii* isolated from *C. compressa*, for example, was two orders higher in the presence of these vitamins than in their absence. The data suggest that these endosymbiotic algae may be receiving vitamins derived from bacteria or other algae that are included in the diets of their hosts.

Several lines of evidence suggest that the diatom-bearing larger foraminifera as symbiotic systems are better adapted to function at light levels that are considerably attenuated than those incident on the surface of the waters where they are found. Growth of *H. depressa* was inhibited at light intensities of 1,200 and 2,400 lx, but the animal whose normal habitat includes shaded sides of tide pools grew well at light intensities between 150 and 600 lx (Röttger 1976). One Hawaiian clone of the animal grew better at 300 lx (Röttger & Berger 1972) than at higher levels tested. Optimum light intensity for the growth of a Hawaiian clone of *A. lessonii* in the laboratory was between 400 and 800 lx (Röttger et al., 1980).

Respirometric studies of the effect of light on primary production of two species of larger foraminifera, *A. hemprichii* and *A. lobifera*, measured in situ in the Red Sea suggested that the algal-animal symbiont systems were photoinhibited at high light intensities and were quite productive $\gtrapprox$10% incident noon light levels (Lee et al. 1980a). Radionuclide tracer experiments gave similar results (Erez 1978).

Experiments on the photosynthetic abilities of axenic clones of endosymbiotic algae from the same hosts suggest that the algae fix more carbon at 312 $\mu W/cm^2$ than they do at 625 $\mu W/cm^2$ (Lee et al., 1980, M. J. Lee et al. 1982). More detailed comparative studies of *F. shiloi, N. laevis, N. valdestriata*, and *N. panduriformis* show greater concentrations of chlorophylls *a* and *c* and carotenoids at light levels approaching those at 30–40 m depth at Elat (9 $\mu W/cm^2$) than at the light levels found at 1 m (280 $\mu W/cm^2$) (M. J. Lee et al., 1982). The growth of *N. valdestriata* [r (intrinsic rate of increase) = 0.383; G (generation time) = 1.27 days] was moderately slow at light levels equal to those found at 1 m in Elat. It grew better (r = 0.545; G = 1.27 days) at light levels approaching those found at 30 m. *N. panduriformis* also grew most rapidly (r = 0.380; G = 1.89 days) at the light levels found at 30 m. *F. shiloi* and *N. laevis* grew most rapidly at the high light levels tested (r = 0.450, G = 1.55 days and r = 0.50, G = 1.33 days, respectively), but growth rates at lower light levels (e.g., 30 m) were not far behind (e.g., r = 0.314, G = 2.21 days and r = 4.17, G = 1.66 days, respectively). Photocompensation values for the four algal species were fairly close together in

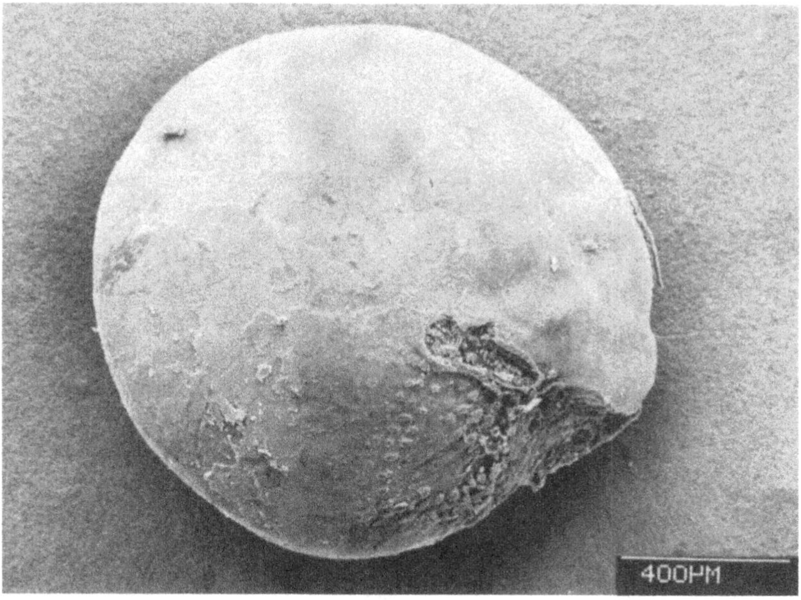

Fig. 14. SEM photograph of *A. lessonii*. Pores on surface barely visible. 60×.

the range of 0.1–0.01 μW/cm² or approximately 0.2%–0.5% of the illumination found at 1 m. If one were to assume moderate light attenuation by the foraminiferal tests then the depth maxima for functioning foram-algal symbiotic systems at Elat would be approximately 40–50 m (light data from Dr. Yehuda Cohen; M. J. Lee et al. 1982).

Morphological adaptations of the host

Studies of the wall structure and morphology of the shells of symbiont-bearing large foraminifera have increased our understanding of the functional adaptations of these animals for symbiosis. One of the more obvious adaptations is observed in the modern nummulites and amphisteginids. The lateral walls of the shell are perforated by numerous parallel pores that extend from the surface of the protoplast to the exterior surface of the shell (Fig. 14). The inner surface of the shell surrounding the pore is pitted or vaulted to form little cups. If turned upside down the inner surface resembles the containers used to protect eggs for sale and transport (Figs. 15, 16). The symbionts are concentrated along the inner surfaces of the chambers in protoplasmic bulges that fit into the pore rim vaults (Fig. 17; Hansen & Buchardt 1977). Though the pores are covered by a thick organic lining (Fig. 15) they are physiologically active. Experiments have shown uptake of neutral red (Berthold 1976) and

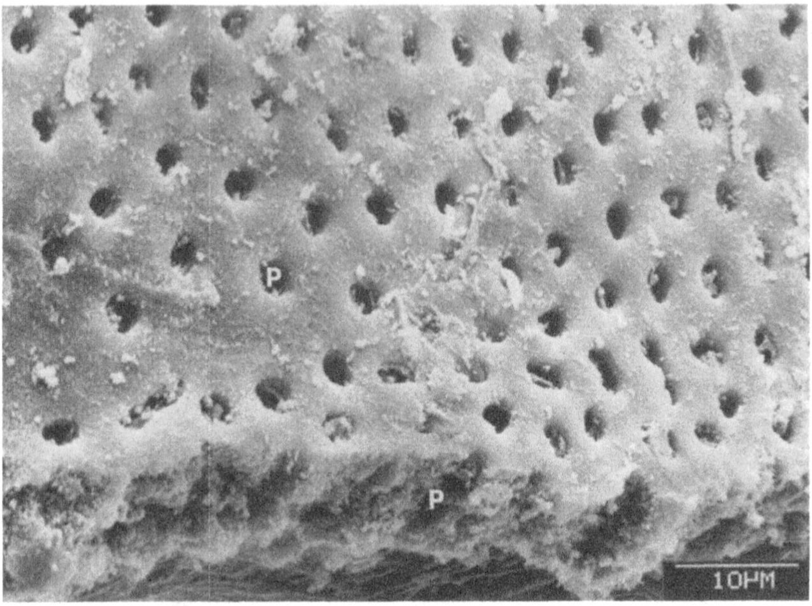

Fig. 15. SEM photograph of *A. lessonii* showing pores. Shell has been fractured (lower part of picture) to show that pores are cylindrical and penetrate almost completely through shell. P, pore. 1,700×.

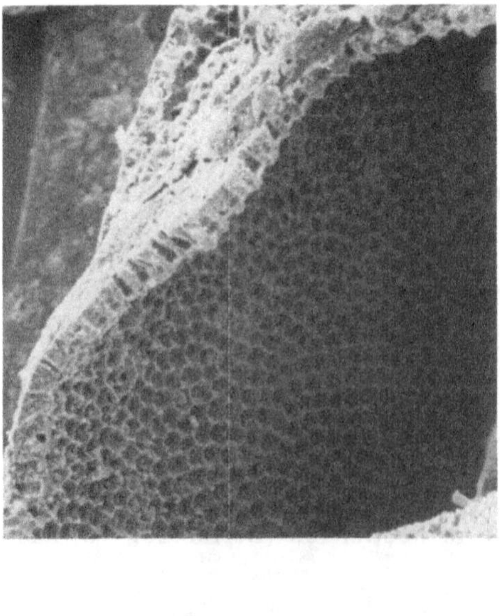

Fig. 16. SEM photograph showing a portion of a shell of *A. lessonii* fractured to reveal the pores and cuplike pore rims on the inner surface of the shell. 700×.

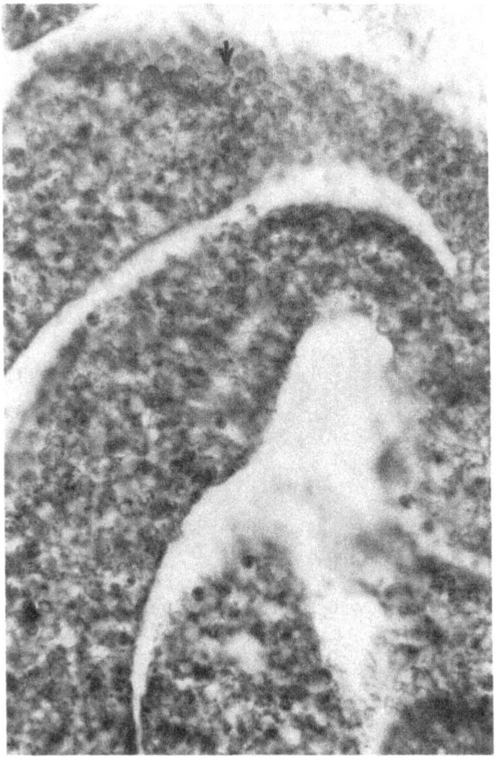

Fig. 17. TLM tangential section of a specimen of *A. lessonii* showing the endosymbiotic diatoms at the outer surface of the cytoplasm directly under the shell. 850×.

$^{14}CO_2$ (Leutenegger & Hansen 1979) through the pores. Mitochondria are concentrated below the pores of nonsymbiont-bearing species, implying physiological activity at these sites (Berthold 1976; Leutenegger & Hansen 1979). The implication that pores facilitate photosynthetic activities by enhancing CO_2 uptake is attractive in the absence of evidence to the contrary. Recent TEM and SEM examinations of the exterior surfaces of imperforate symbiont-bearing species such as *Marginopora vertebralis, Cyclorbiculina compressa, S. orbiculus, Peneroplis pertusus*, and *Spirolinata arietina* have shown that, although these animals do not have pores, their surfaces are pitted or ribbed and pitted (e.g., Fig. 18) (Hansen & Dalberg 1979). Hansen and Dalberg (1979) also interpret these features as adaptations to minimize lateral wall thickness and facilitate passage of CO_2 to their symbionts.

The overall shape and construction of symbiont-bearing foraminifera have also been interpreted as a compromise between hydrodynamic factors and maximizing surface area to expose algae to light (Hallock 1979, Haynes 1965). Many of the symbiont-bearing forms are disk shaped with thin outer chamber walls. The shells are strengthened by internal pillars,

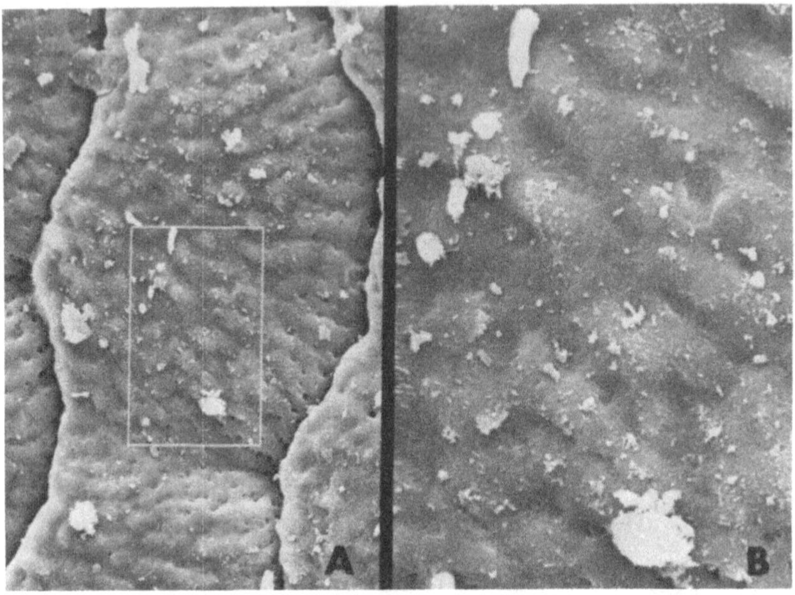

Fig. 18. SEM of the surface of *A. hemprichii* showing ribbed **(A)** and pitted **(B)** surface. **A** 680×; **B** 1,200×.

laminae of calcite, chamberlet walls, and other similar structures designed to maintain the structural integrity of the test while maximizing light penetration (Hottinger 1978). The roller-shaped fusilines and alveolines in the fossil record had thin outer walls that are now reasonably interpreted as adaptations for algal endosymbiosis. A number of workers have commented on the differences in the shape of symbiont-bearing foraminifera collected at different depths (Hallock 1979, Hottinger 1977, Hottinger & Dreher 1974, Larsen 1976, Larsen & Drooger 1977). Since there is no thermocline in the clear waters of the Gulf of Elat, where many of the foregoing studies were carried out, Larsen (1976) suggested that the increase in surface to volume with depth observed in *Amphistegina* was causally linked to the light gradient. Hallock and Hansen (1979) recently obtained additional evidence to support Larsen's hypothesis. They found that the secondary lamellae were thinned with depth, causing corresponding changes in the test shape. Hallock (1979) has also shown that laboratory clones of *A. lessonii* and *A. lobifera* produced thicker tests when grown at high light levels than when grown in reduced light.

Behavioral adaptations of algal-foraminiferan symbioses

The results of several very simple experiments on phototaxis in several symbiont-bearing species indicate that they have a behavioral repertoire that includes responses to photic stimuli (Lee et al. 1980a, Zmiri et al. 1974). *Amphistegina radiata* was unresponsive at low light levels but was positively phototactic between photonic fluxes of 10^{11} – 10^{15} photons/ cm^2 per second. The peak of the action spectrum for this response was near 500 nm (Zmiri et al. 1974). In another set of experiments (Lee et al., 1980a), *A. lobifera* was negatively phototactic at an incident illumination of 10 klx and positively phototactic between 1 and 0.1 klx. *A. hemprichii* was positively phototactic between 10 and 20 klx to <6 klx but $>$ 1.7 klx. The positive phototactic response in *A. hemprichii* was stronger than their feeding territorial response since many individuals sacrificed their territoriality to move either toward or away from light in response to a particular light regime.

A fascinating behavioral response to algal symbiosis has been observed in several planktonic species, *Orbulina universa, Globigerinoides ruber, G. sacculifer*, and *G. conglobatulus*. Anderson, Bé and co-workers (Anderson and Bé 1976b, Bé and Hutson 1977, Bé et al. 1977) have observed circadian periodicity in which nonmotile dinoflagellate zooxanthellae emerge in the pseudopods at dawn from the shell and are moved to the distal parts of the spines and rhizopodial net and then are drawn through the primary and secondary apertures or, in the case of *O. universa*, into the larger pores in the evening. This adaptation presumably maximizes the exposure of the algae to light and facilitates nutrient uptake. Light seems to be the triggering mechanism for migration since the zooxanthellae will emerge from the shell upon exposure to microscope illumination after the animals have been kept in the dark for several hours (Bé et al. 1977).

The trophic significance of endosymbiosis

Several workers have attempted to evaluate the value of the symbiotic relationship to the overall nutrition of the host animal. A culture approach has been used by workers in Kiel, Germany (Röttger 1972, Röttger et al. 1980) in their studies of the effect of light on the growth of Hawaiian clones of *H. depressa* and *A. lessonii*. Within the range tested, growth of both species was proportional to light intensity. *A. lessonii* grew best at 800 lx, whereas different clones of *H. depressa* grew best between 400 and 600 lx. Neither species grew in the dark. In one experiment they kept *A. lessonii* in the dark for 77 days, offering it a diet of autoclaved mashed *Cladophora socialis*, detritus of unicellular algae, bacteria, protozoa, and fungi or yeast, but the animals did not grow. Controls incubated in the light grew normally, leading the investigators to conclude that the growth of the two symbiont-bearing species depends mainly on the photosynthesis of their symbiotic diatoms. Evidence obtained from ^{14}C tracer studies of primary production in *A. lessonii* tends to support the findings of the Kiel group (Muller 1978). Individual animals fixed 1 x 10^{-5} μCi $NaH^{14}CO_3$ h^{-1}. In a light regime of 12 h light/12 h dark the biological half-life of the tracer seemed to be ~ 50 h. Deposition into the calcium carbonate skeleton accounted for only 10% of the total uptake. By manipulating the light and dark periods and estimating the rates of loss of the ^{14}C label, Muller (1978) concluded that the foraminifera and their symbionts recycle about half of their respired carbon and that the animal depends upon its algal symbionts for growth and carbonate production.

In situ, respirometric and radionuclide tracer feeding techniques have also been used to estimate symbiont contributions to host trophic dynamics. Glass differential respirometers mounted on floats were used to measure primary production (Lee & Bock 1976, Lee et al., 1980a). The algae used in the feeding experiments were clones that were either found to be excellent food organisms for a variety of benthic foraminifera or were abundant in the habitat where the foraminifera were collected. Data from Key Largo Sound on two species, *S. marginalis* and *A. angulatus*, clearly indicated that at midday feeding was the most important process; the ratio of carbon gain in both animals by feeding to primary production was ≧ 10:1. Both animals were selective feeders, eating large quantities of different algal species. Their diets were relatively nonoverlapping. Using these methods Lee and co-workers (1980a) found similar results in two of the three species they studied in situ in the Red Sea. They starved their animals in the light for 1 week before the experiment, hoping to exaggerate feeding, if possible. *A. lobifera* and *A. hemprichii* were selective feeders. The ratio of carbon uptake through feeding to that fixed by the symbionts was ~ 10:1. Under the same experimental conditions *H. depressa* ate only small numbers of algae. There was some suggestion from a feeding experiment that used an agnotobiotic natural mixture that this animal might eat some bacteria. These data are consistent with Röttger's (1976) evidence that almost the entire carbon budget of this species of larger foraminifera is satisfied by their endosymbiotic diatoms.

The recent study of feeding in the symbiont-bearing planktonic foraminifer *G. sacculifer* in laboratory culture by Bé and his co-workers (1981) also suggests that the main source of nutrition for this animal must be active feeding. Symbiont photosynthetic activities were insufficient as a source of nutrition since starved animals grew slowly, if at all, and most failed to reproduce.

E. williamsoni collected in a Danish brackish habitat, Limfjorden, contained on the average 4×10^6 chloroplasts mg^{-1} (Fig. 19). *N. germanicum* contained about 10% fewer chloroplasts (Lopez 1979). The chloroplasts are apparently functional since $H^{14}CO_3^-$ uptake in the light was proportional to chlorophyll content and light intensity and was insignificant

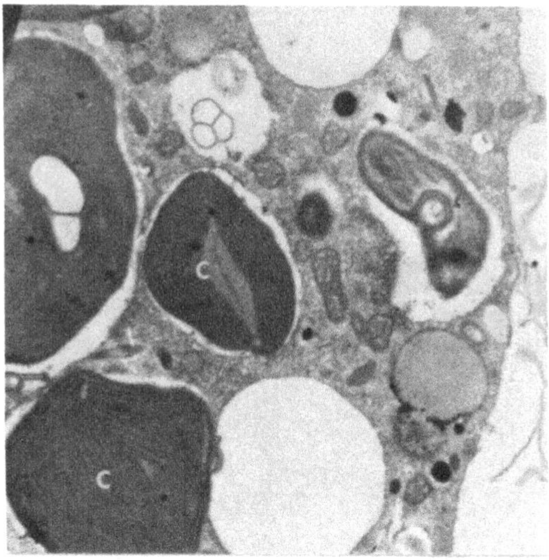

Fig. 19. TEM micrograph of a small section of *E. williamsoni* showing chloroplasts. Photograph by Ellen Lopez. 5,600×.

in the dark. At light saturation (10 klx) the chloroplasts in *E. williamsoni* assimilated 2.3×10^{-3} mgC mg^{-1} free dry weight h^{-1}. Chloroplasts survived longer in foraminiferans kept in the dark than those adapted to a light/dark regime. Individual *E. williamsoni* must eat at least 65 chloroplasts per hour, and *N. germanicum* at least 20, to maintain a steady state population of chloroplasts. Additional details of this interesting symbiotic association are being studied by Lopez (personal communication) in order to evaluate its nutritional and shell-building significance.

Cytological and fine structural studies of animals fixed freshly in the field tend to support the concept that most larger foram species actively feed on algae (e.g., Leutenegger 1977a, b, McEnery & Lee 1981, Müller-Merz & Lee 1976). An exception seems to be *H. depressa*. Few food vacuoles or remnants of digested diatoms have been encountered in the intratest protoplasm of this animal. The endosymbiotic algae in *H. depressa* are fairly uniformly distributed throughout their host but tend to be more fusiform in outer chambers (Fig. 20) where digestion of algae takes place in most disk-shaped larger foraminifera (McEnery & Lee 1981). It has been suggested that this animal may feed by digesting some of its zooxanthellae (Dietz-Elbrachter 1971). There is no experimental evidence to support this idea but it has not yet been carefully researched.

Even though larger foraminifera are unicellular, their cytoplasm is quite compartmentalized. Perhaps as in the case of *Paramecium bursaria* (Weis 1976) and green *Hydra* (Pardy & Muscatine 1973) this is a mechanism to segregate algal endosymbionts from digestive activities. As noted earlier in this chapter, the algal endosymbionts in nummulites and amphistegines tend to be segregated to the upper and lateral walls of the

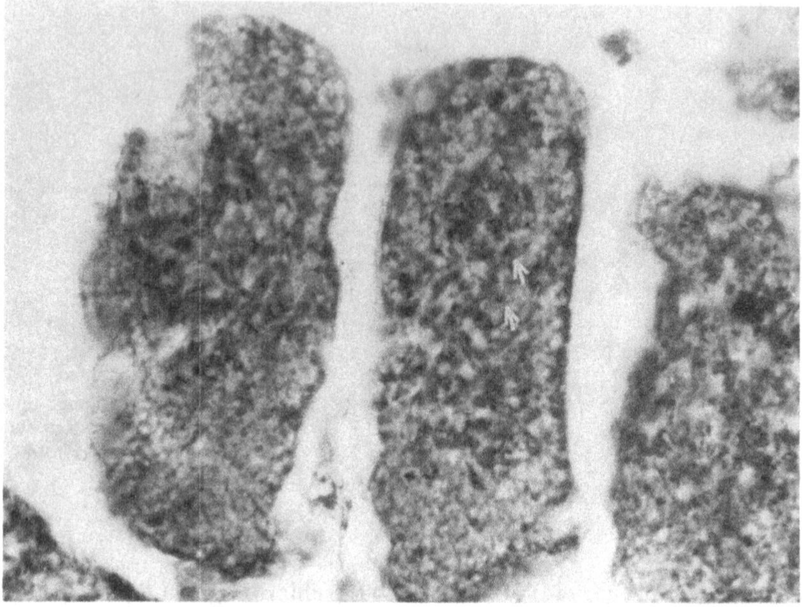

Fig. 20. TLM equatorial section of several chamberlets of *H. depressa* showing the more elongated *N. panduriformis* var. *continua* in the cytoplasm. 950×.

animals. In *Amphistegina* spp. food vacuoles tend to be located toward the center of the ventral region. Diatoms in digestive vacuoles clearly have frustules.

It was possible to observe valve views of many diatoms in food vacuoles and identify them as to genus or species (McEnery & Lee 1981). In no instance were the diatoms in food vacuoles identified as belonging to the same species as any presently known endosymbiotic diatom species.

An extreme in cytological differentiation was found in one of the coin-shaped soritids, *S. marginalis*, collected in Key Largo Sound (Müller-Merz & Lee 1976). The endosymbiont in this species, *S. microadriaticum*, tends to be packed in the intermediate chambers. Very few are found in the embryonic inner, or outer chambers. Food vacuoles were found only in the outer chambers. Some of the zooxanthellae in the outer chambers near food vacuoles seemed moribund or undergoing digestion. This foraminiferan host is heterokaryotic, with hundreds of generative (micro-) nuclei and scores of vegetative (somatic, macro-) nuclei. Most of the generative nuclei were restricted to the embryonic chambers whereas the vegetative nuclei were most abundant in the intermediate and outer chambers among the symbionts and digestive vacuoles. The cellular organization of *A. hemprichii* was similar to *S. marginalis* but not quite as highly regionalized (McEnery & Lee 1981). Food vacuoles are

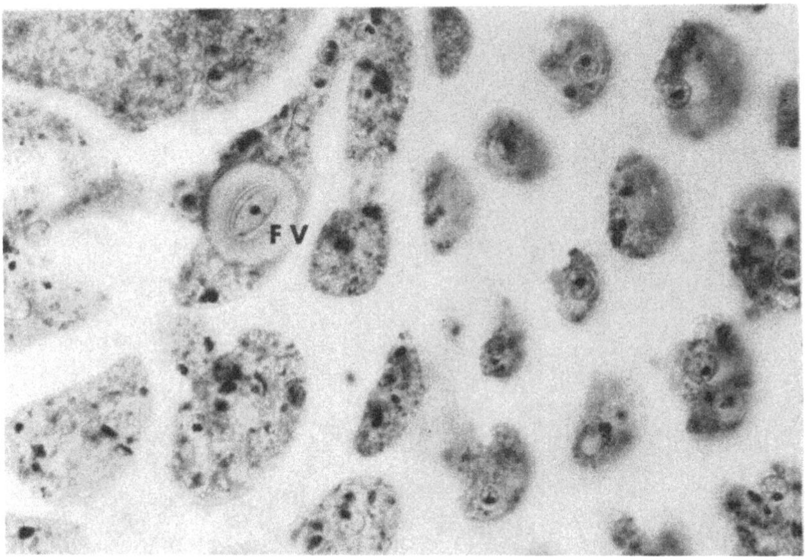

Fig. 21. TLM equatorial section of *A. hemprichii* showing outer chamberlets. Note the food vacuole (FV) with pennate diatom at center right of photograph. 950×.

more widely distributed throughout the animal but concentrated more heavily in the outer chambers (Fig. 21). As in *S. marginalis* the zooxanthellae in *A. hemprichii* are more concentrated in the intermediate chambers (Fig. 22) where they occupy ~80% of the total chamber volume, but they are more widely abundant in other chambers than they are in *S. marginalis*.

The planktonic foraminifer *G. sacculifer* possesses at least three cytologically distinct types of zooxanthellae (Anderson & Bé 1976b). The most numerous one, a small species (5 μm × 8 μm) of *Aureodinium*, occupies almost 50% of the chambers into which they are withdrawn at night. Bé and Hutson (1977) estimate that there are several hundred of these algae per animal.

Cytological and fine structural studies hint that in some cases algal-fixed carbon may be passed to the host after the death of endosymbionts by autolysis or digestion. Certainly many of the zooxanthellae in *S. marginalis* fixed in situ were filled with starch almost to the point of bursting (Figs. 5, 23). The cytoplasm of the host contained similar-sized starch grains. One can imagine that this starch might be an important nutritional source for the animals if its release from burst cells into the cytoplasm occurs with any regularity. Only a few studies have attempted to evaluate the potential for transfer of metabolites from photosynthesizing algae in larger foraminifera to their hosts. Though it is not clear from their discussion of materials and methods how long they incubated cultures in a

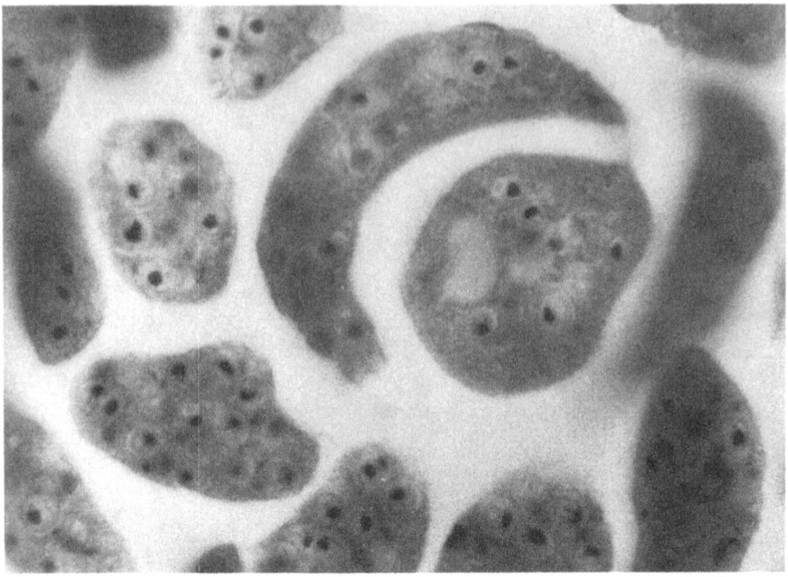

Fig. 22. TLM equatorial section of *A. hemprichii* through embryonic apparatus and inner chambers showing the abundance of dinoflagellate zooxanthellae. 1,150×.

$H^{14}CO_3^-$ tracer or why they chose a light level of 5,000 lx for this particular study, Kremer et al. (1980) used chromatographic techniques to identify the major photosynthates in algal-host extracts of six species of larger foraminifera incubated in the laboratory. They identified floridoside (2-0-D glycerol-D galactopyranoside) and polyglucan in extracts of *Spirolina arietina* and *Peneroplis pertusus*, two rhodophycean-bearing foraminifera. Of the labeled photosynthate produced by the dinoflagellates in *A. hemprichii*, 74% was unspecified lipids. Another 3.5% was glycerol. A large percent of the label, 31%, 51%, and 33%, respectively, was also found in the unspecified lipids of the diatom-bearing species *A. lessonii, A. lobifera*, and *H. depressa*. An additional 5%, 6%, and 11%, respectively, of the label was found in glycerol. Polyglucan was the largest labeled fraction (45% and 43%) in *A. lobifera*. Although we as reviewers are skeptical because of the technical approaches used in the published report, Kremer and his associates (1980) concluded that their results demonstrate that the animals receive organic carbon by release of algal assimilates rather than by digestion of their algal symbionts.

Although it has been demonstrated that release of metabolites is stimulated by host tissues in a variety of symbiotic associations (Smith 1975, and reviewed in Trench 1980), one foraminiferan endosymbiont, *C. hedleyii* from *A. angulatus*, releases large quantities of metabolite in axenic culture (Lee et al. 1974). After 10 days' incubation (end of log phase)

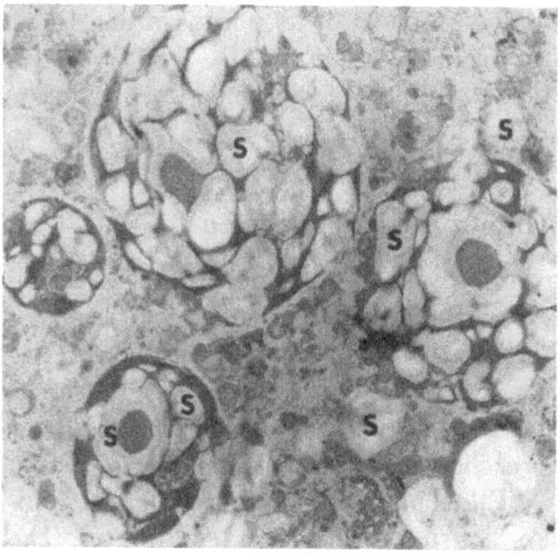

Fig. 23. TEM section of a chamberlet of *Cyclorbiculina compressa* showing *C. provasolii* filled with starch grains (S) and some starch grains in the cytoplasm of the host. From a study by Edith Müller-Merz and John J. Lee (1976). *Top.* 4,350×; *Bottom.* 7,680×.

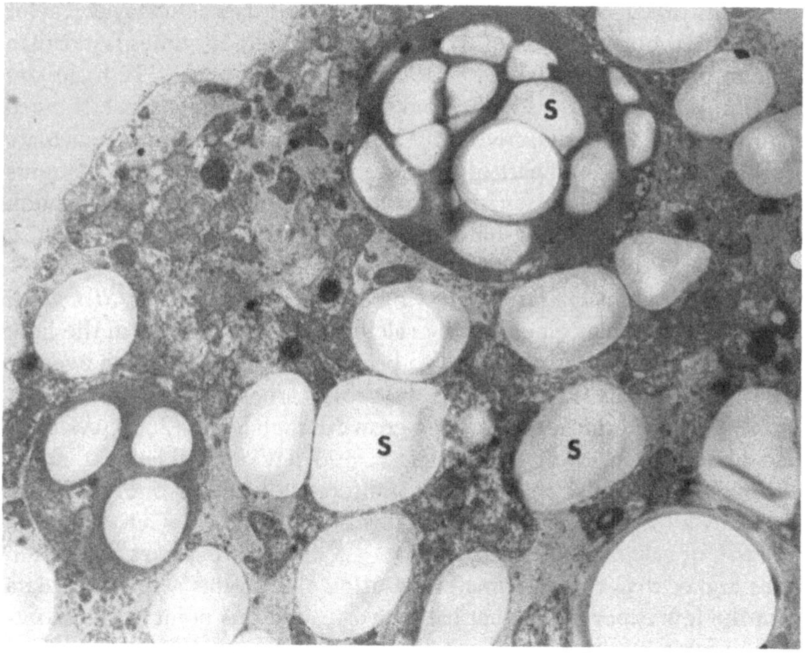

in $NaH^{14}CO_3$ tracer-labeled mineral medium, they found more labeled organic carbon in the medium (57%) than they did in the cells. The organic material in the medium was chromatographically homogeneous and they tentatively identified it as mannitol. Saks (personal communication) is currently studying the release of metabolites from axenic cultures of *C. provasolii*. Under his test conditions, in contrast with *C. hedleyi*,

C. provasolii seems to release only 7% or so of its photosynthates into the medium. Two *Chlorella* spp. from *A. lessonii* and *A. hemprichii* released only 1% of their carbon fixed to their growth medium as glucose and mannose (Saks 1981).

Role of endosymbionts in calcification

The role of endosymbionts in the growth and calcification of foraminiferan tests (shells) is much clearer. Photosynthesis and calcification, as measured with a ^{45}Ca tracer in *A. angulatus*, were shown to be directly proportional to light intensity in the range of 0–200 Einstein's $\mu E \cdot m^{-2} \cdot s^{-1}$ and two to three times that observed in the dark (Duguay & Taylor 1978). Saturation was reached near 200 Einstein's $\mu E \cdot m^{-2} \cdot s^{-1}$. The herbicide DCMU [3-(3,4-dichlorophenyl)-1,1-dimethylurea] completely inhibited photosynthesis and light enhanced calcification. The rate of calcification on the basis of dry weight was faster in younger specimens than in older ones. According to their data (Duguay & Taylor 1978), a 0.5-mm specimen of *A. angulatus* calcified 1.5 times faster than a 2-mm one. Smith and Wiebe (1977), in a "long-term" (91 h) in situ study of *Marginopora vertebralis* from Lizard Island, Queensland, showed that ~20% of the radioactivity (^{14}C) fixed in the light into particulate organic carbon became part of the calcareous shell. Negligible amounts were incorporated in the dark or in formalin-killed controls. Calcification rates of *A. lobifera* in the Gulf of Elat ranged from 938 to 3,481 μg Ca/g $CaCO_3$ h (Erez 1978). This rate is five times higher than the average calcification rate of the fastest growing coral in the same reef, *Stylophora pistillata*. The enhancement of the calcification of *A. lobifera* in the light over dark controls was estimated to be ~50 × compared to an average of 11 × for corals (Erez 1978). Laboratory clones of *A. lessonii* and *A. lobifera* produced thicker tests when grown at high levels (2600 $\mu W/cm^2$) than under much lower light levels (~300 $\mu W/cm^2$) (Hallock 1979). Although ideally it should make no difference whether one uses ^{45}Ca or ^{14}C as a tracer to estimate the rates of building a $CaCO_3$ skeleton, the rates of calcification estimated by ^{45}Ca uptake are always five to ten times higher than those estimated by using ^{14}C. In this respect the data from the few experiments that have investigated this point in symbiont-bearing foraminifera are quite in agreement with data from corals (Erez 1978).

Stable isotope disequilibrium studies also provide evidence that symbiotic algae produce a light-dependent vital effect on the shells of foraminifera that bear them (Buchardt & Hansen 1977, Erez 1978). The isotopic composition of the shell ($^{13}C/^{12}C$; $^{18}O/^{16}O$) becomes lighter when the rate of photosynthesis increases. In the Gulf of Elat there was an average $\delta\ ^{18}O$ for symbiont-bearing species 1.5% lighter than the expected equilibrium value (Buchardt & Hansen 1977, Erez 1978). Controls were mollusc shells and nonsymbiont-bearing foraminifera in isotopic equilibrium with their surrounding seawater. Oxygen and carbon isotope analyses have also been made on symbiont-bearing benthic foraminifera from Bermuda, the Persian Gulf, and the Philippines. The miliolids studied, *M. vertebralis, Cyclorbiculina compressa, A. angulatus, Peneroplis proteus*, and *Praesorites orbitolitoides*, have carbon isotope values more than 2.5% lighter than expected equilibrium values for their sample localities. The shells of the rotaliids *H. depressa, Operculina* sp. and *Calcarina spengleri* were similar ($\delta\ ^{13}C \approx 2\%$ lighter) (Weter et al. 1981). Since isotope disequilibrium has also been observed in many taxa of recent nonsymbiont-bearing deep-sea foraminifera, additional interest has been piqued in the study of vital effects and how they might affect the accuracy of biogenic carbonate estimations of paleotemperatures, depth habits, etc. (e.g., Ahmad & Perry 1980, Woodruff et al. 1980).

Unresolved problems

Although many aspects of the biology of symbiosis are unresolved, several areas are particularly ripe for study. Among these are (1) the relationship(s) between photosynthesis, shell carbonate production, and isotope disequilibrium; (2) the specificity and requirements of host foraminifera and of algae capable of being endosymbionts; and (3) the nutritional carbon transfer (if any) from the algae to the animal. Since research in these problem areas is technically feasible we look forward to considerable progress in the next few years.

References

Ahmad, S., & Perry, E. Jr. (1980). Isotopic evolution of the sea. *Scientific Progress* **66**, 499–511.

Alldredge, A. L., & Jones, B. M. (1973). *Hastigerina pelagica*: foraminiferal habitat for planktonic dinoflagellates. *Mar. Bio.* **22**, 131–315.

Anderson, O. R., & Bé, A. (1976a). A cytochemical fine structure study of phagotrophy in a planktonic foraminifer, *Hastigerina pelagica* (d'Orbigny). *Biol. Bull.* **151**, 437–49.

– (1976b). The ultrastructure of a planktonic foraminifer, *Globigerinoides sacullifer* (Brady), and its symbiotic dinoglagellates. *J. Foram. Res.* **6**, 1–21.

Bé, A., & Hutson, W. H. (1977). Ecology of planktonic and biogeographic patterns of life and fossil assemblages in the Indian Ocean. *Micropaleontology* **23**, 369–14.

Bé, A., Hemleben, D., Anderson, O., Spindler, M., Hacunda, J., & Tuntivate-Choy, S. (1977). Laboratory and field observations of living planktonic foraminifera. *Micropaleontology* **23**, 155–79.

Bé, A., Caron, D. A., & Anderson, O. R. (1981). Effects of feeding frequency on life processes of the planktonic foraminifer *Globigerinoides sacculifer* in laboratory culture. *J. Mar. Bio. Ass. U. K.* **61**, 257–77.

Berthold, W. (1976). Ultrastructure and function of wall perforations in *Patellina corrugata* Williamson, foraminifera. *J. Foram. Res.* **6**, 22–9.
– (1978). Ultrastruktur-analyse der endoplasmatischen Algaen von *Amphistegina lessonii* d'Orbigny, Foraminifera (Protozoa) und ihre systematische Stellung. *Archiv für Protistenk.* **120**, 16–62.
Brasier, M. D. (1975a). Ecology of recent sediment-dwelling and phytal foraminifera from the Lagoon of Barbuda, West Indies. *J. Foram. Res.* **5**, 42–62.
– (1975b). Morphology and habitat of living Benthonic foraminiferids from Caribbean carbonate environments. *Revista Espanola De Micropaleontologia* **7**, 567–78.
Buchardt, B., & Hansen, H. (1977). Oxygen isotope fractionation and algal symbiosis in benthic foraminifera from the Gulf of Elat, Israel. *Bull. Geol. Soc. Denmark* **26**, 185–94.
Butschli, O. (1886). Beitrage zur Kenntnis des Flagellaten und einiger verwundter Organismen. *Morphologisches Jahrbeitung* **11**, 78–101.
Buzas, M. (1968). On the spatial distribution of foraminifera. *Contributions for the Cushman Foundation of Foraminiferal Research* **19**, 1–11.
– (1969). Foraminiferal species densities and environmental variables in an estuary. *Limnol. Oceanog.* **14**, 411–22.
– (1970). Spatial homogeneity: Statistical analyses of unispecies and multispecies populations of foraminifera. *Ecology* **51**, 874–9.
Cushman, J. A. (1922). Shallow-water foraminifera of the Tortugas region. *Papers from the Tortugas Laboratories. Carnegie Institute of Washington* **32**, 127–42.
Dietz-Elbrachter, G. (1971). Untersuchungen über die zooxanthellen der Foraminifera *Heterostegina depressa* d'Orbigny 1826. *Meteor Forschungsgemeinschaft Ergebnisse* **6**, 41–7.
Duguay, L., & Taylor, D. (1978). Primary production and calcification by the soritid foraminifera *Archaias angulatus* (Fichtel and Moll). *J. Protozoo.* **25**, 356–61.
Erez, J. (1978). Vital effect on stable-isotope composition seen in foraminifera and coral skeletons. *Nature* **273**, 199–202.
Hallock, P. (1979). Trends in test shape with depth in large symbiont-bearing foraminifera. *J. Foram. Res.* **9**, 61–9.
Hallock, P., & Hansen, H. (1979). Depth adaptation in *Amphistegina*: Change in lamellar thickness. *Bull. Geo. Soc. Denmark* **27**, 99–104.
Hansen, H., & Buchardt, B. (1977). Depth distribution of *Amphistegina* in the Gulf of Elat, Israel. *Utrecht Micropaleontological Bulletins* **15**, 205–44.
Hansen, H., & Dalberg, P. (1979). Symbiotic algae in milioline foraminifera: CO_2 uptake and shell adaptations. *Bull. Geo. Soc. Denmark* **28**, 47–55.
Haynes, J. (1965). Symbiosis, wall structure and habitat in foraminifera. *Contributions from the Cushman Foundation for Foraminiferal Research* **16**, 40–3.
Hottinger, L. (1977). Distribution of larger Peneroplidae, *Borelis* and Nummulitidae in the Gulf of Elat, Red Sea. *Utrecht Micropaleontological Bulletins* **15**, 35–109.
Hottinger, L. (1978). Comparative anatomy of shell structures in selected larger foraminifera. *In: Foraminifera*, vol. 3, ed. R. H. Hedley and C. G. Adams, London: Academic Press.
Hottinger, L., & Dreher, D. (1974). Differentiation of protoplasm in Nummulitidae (Foraminifera) from Elat, Red Sea. *Mar. Bio.* **25**, 41–61.
Kessler, E. (1976). Comparative physiology, biochemistry, and the taxonomy of *Chlorella* (Chlorophyceae). *Plant Syst. Evol.* **125**, 129–38.

Kessler, E. (1978). Physiological and biochemical contributions to the taxonomy of the genus *Chlorella*. XII. Starch hydrolosis and a key for the identification of 13 species. *Arch. Microbiol.* **19**, 13–16.
Kessler, E., & Czygan, F-C. (1970). Physiological and biochemical contributions to the taxonomy of the genus *Chlorella*. IV. Utilization of organic nitrogen compounds. *Arch. Microbiol.* **70**, 211–16.
Kessler, E., & Zweier, I. (1971). Physiological and biochemical contributions to the taxonomy of genus *Chlorella*. V. Auxotrophic and mesotrophic species. *Archiv. Microbiol.* **79**, 44–8.
Kremer, B., Schmaljohann, R., & Röttger, R. (1980). Features and nutritional significance of photosynthetates produced by unicellular algae symbiotic with larger foraminifera. *Marine Ecology Progress Series* **2**, 225–8.
Larsen, A. R. (1976). Studies on recent *Amphistegina*, taxonomy and some ecological aspects. *Israel J. Earth Sci.* **25**, 1–26.
Larsen, A. R., & Drooger, C. W. (1977). Relative thickness of the test in the *Amphistegina* species of the Gulf of Elat. *Utrecht Micropaleontological Bulletins* **15**, 225–40.
Lee, J. (1974). Towards understanding the niche of foraminifera. *In: Foraminifera*, vol. 1, ed. R. H. Hedley & C. G. Adams, pp. 208–57. London: Academic Press.
– (1980). Nutrition and physiology of the foraminifera. *In: Biochemistry and Physiology of Protozoa*, end ed., vol. 3, ed. M. Levandowsky & S. H. Hutner, pp. 43–66. New York: Academic Press.
Lee, J., & Bock, W. (1976). The importance of feeding in two species of soritid foraminifera with algal symbionts. *Bull. Mar. Sci.* **26**, 530–7.
Lee, J. J., & Reimer, C. W. (in press). Isolation and identification of endosymbiotic diatoms from larger foraminifera of the Great Barrier Reef, Australia, Makapuu Tide Pool, Oahu, Hawaii, and the Gulf of Elat, Israel with the descriptions of three new species *Amphora rottgerii, Navicula lansenii*, and *Nitzschia frustrulum* variety symbiotica.
Lee, J., & Zucker, W. (1969). Algal flagellate symbiosis in the foraminifer *Archaias*. *J. Protozoo.* **16**, 71–81.
Lee, J., Freudenthal, H., Kossoy, V., & Bé, A. (1965). Cytological observations on two planktonic foraminifera, *Globigerina bulloides* d'Orbigny, 1826, and *Globigerinoides ruber* (d'Orbigny, 1839), Cushman, 1927. *J. Protozoo.* **12**, 531–42.
Lee, J., McEnery, M., & Rubin, H. (1969). Quantitative studies on the growth of *Allogromia laticollaris* (Foraminifera). *J. Protozoo.* **16**, 377–95.
Lee, J., Crockett, L., Hagen, J., & Stone, R. (1974). The taxonomic identity and physiological ecology of *Chlamydomonas hedleyi*, sp. nov., algal flagellate symbiont from the foraminifer *Archaias angulatus*. *Bri. J. Phycol.* **9**, 407–22.
Lee, J., McEnery, M., Kahn, E., & Schuster, F. (1979). Symbiosis and the evolution of larger foraminifera. *Micropaleontology* **25**, 118–40.
Lee, J., McEnery, M., & Garrison, J. (1980a). Experimental studies of larger foraminifera and their symbionts from the Gulf of Elat on the Red Sea. *J. Foram. Res.* **10**, 31–47.
Lee, J., Reimer, C., & McEnery, M. (1980b). The identification of diatoms isolated as endosymbionts from larger foraminifera from the Gulf of Elat (Red Sea) and the description of 2 new species, *Fragilaria shiloi* sp. nov. and *Navicula reissii* sp. nov. *Bot. Mar.* **23**, 41–8.
Lee, J., McEnery, M., Röttger, R., & Reimer, C. (1980c). The isolation, culture and identification of endosymbiotic diatoms from *Heterostegina depressa*

d'Orbigny and *Amphistegina lessonii* d'Orbigny (larger foraminifera) from Hawaii. *Bot. Mar.* **23**, 297–302.

Lee, J., McEnery, M., Lee, M., Reidy, J., Garrison, J., & Röttger, R. (1980d). Algal symbionts in larger foraminifera. *In: Endocytobiology*, vol. 1, ed. W. Schwemmler & H. E. A. Schenk, pp. 113–24. Berlin: Walter de Gruyter & Co.

Lee, J. J., J. Reidy, and E. Kessler (1982). Symbiotic *Chlorella* species from larger foraminifera. *Bot. Mar.* **25**, 171–6.

Lee, M. J., R. Ellis, and J. J. Lee. (1982). A comparative study of photoadaption in four diatoms isolated as endosymbionts from larger foraminifera. *Mar. Biol.* **68**, 193–7.

Leutenegger, S. (1977a). Symbiosis between larger foraminifera and unicellular algae in the Gulf of Elat. *Utrecht Micropaleontological Bulletins* **15**, 22–239.

– (1977b). Ultrastructure de foraminiferes perfores et imperfores ainsi que de leurs symbiotes. *Cahiers de Micropaleontologie* **3**, 5–53.

– (1977c). Ultrastructure and motility of dinophyceans symbiotic with larger, imperforated foraminifera. *Mar. Bio.* **44**, 157–64.

Leutenegger, S., & Hansen, H. (1979). Ultrastructural and radiotracer studies of pore function in foraminifera. *Mar. Bio.* **54**, 11–16.

Levanson-Spanier, I., Padan, E., & Reiss, Z. (1979). Primary production in a desert-enclosed sea – the Gulf of Elat (Aqaba), Red Sea. *Deep Sea Research* **26/6A**, 673–85.

Loeblich, A., & Sherley, J. (1979). Observations on the theca of the motile phase of free-living and symbiontic isolates of *Zooxanthella microadriatica* (Freudenthal) comb. nov. *J. Mar. Bio. Ass. U. K.* **59**, 195–205.

Lopez, E. (1979). Algal chloroplasts in the protoplasm of three species of benthic foraminifera: Taxonomic affinity, viability and persistence. *Mar. Bio.* **53**, 201–11.

Lutze, G. F. (1980). Habitat and asexual reproduction of *Cyclorbiculina compressa* (d'Orbigny), Soritidae. *J. Foram. Res.* **10**, 251–60.

Lynts, G. W. (1966). Variation of foraminiferal standing crop over short lateral distances in Buttonwood Sound, Florida Bay. *Limnol. Oceanogr.* **11**, 502–66.

McEnery, M., & Lee, J. (1981). Cytological and fine structural studies of 3 species of symbiont-bearing larger foraminifera from the Red Sea. *Micropaleontology* **27**, 71–83.

Matera, N., & Lee, J. (1972). Environmental factors affecting the standing crop of foraminifera in sublittoral and psammolittoral communities of a Long Island salt marsh. *Mar. Bio.* **14**, 89–103.

Muller, P. (1974). Sediment production and population biology of the benthic foraminifer *Amphistegina madagascariensis*. *Limnol. Oceanogr.* **19**, 802–9.

– (1978). Carbon fixation and loss in a foraminiferal-algal symbiont system. *J. Foram. Res.* **8**, 35–41.

Müller-Merz, E., & Lee, J. (1976). Symbiosis in the larger foraminifera *Sorites marginalis* (with notes on *Archaias* sp.). *J. Protozoo.* **23**, 390–6.

Newell, R. (1949). Phyletic size increase, an important trend illustrated by fossil invertebrates. *Evolution* **3**, 103–24.

O'Brien, T. (1978). An ultrastructural study of zoochlorellae in a marine coelenterate. *Trans. Amer. Micros. Soc.* **97**, 320–9.

Pardy, R., & Muscatine, L. (1973). Recognition of symbiotic algae by *Hydra viridis*. A quantitative study of the uptake of living algae by aposymbiotic *H. viridis*. *Bio. Bull.* **145**, 565–79.

Rhumbler, L. (1909). Die Foraminiferen (Thalamophoren) der Plankton-Expedition. I. *Lief c. Ergebnisse der Plantkon-Expedition der Humbolt-Stiftung* **3**, 1–331.
Ross, C. A. (1972). Biology and ecology of *Marginopora vertebralis* (foraminifera) from the Great Barrier Reef. *J. Protozool.* **19**, 181–92.
Ross, C. (1974). Evolutionary and ecological significance of large calcareous Foraminiferida (Protozoa), Great Barrier Reef. *Proc. of the Second Intern. Coral Reef Symposium* 1. Brisbane: Great Barrier Committee.
Röttger, R. (1972). Analayse von Wachstumskurwen von *Heterostegina depressa* (Foraminifera: Nummulitidae). *Mar. Bio.* 17, 228–42.
– (1976). Ecological observations of *Heterostegina depressa* (Foraminifera, Nummulitidae) in the laboratory and in its natural habitat. *Maritime Sediments, Special Publication* **1**, 75–9.
Röttger, R., & Berger, W. (1972). Benthic foraminifera: Morphology and growth. *Mar. Bio.* **15**, 84–9.
Röttger, R., Irwan, A., Schmaljohann, R., & Franzisket, L. (1980). Growth of the symbiont-bearing foraminifer *Amphistegina lessonii* d'Orbigny and *Heterostegina depressa* d'Orbigny (Protozoa). *In: Endocytobiology*, vol. 1, ed. W. Schwemmler & H. E. A. Schenk, pp. 125–32. Berlin: Walter de Gruyter & Co.
Saks, N. M. (1981). Growth, productivity and excretion of *Chlorella* spp. endosymbionts from the Red Sea: Implications for lost foraminifera. *Bot. Mar.* **24**, 445–9.
Schmaljohann, R., & Röttger, R. (1976). Die Symbioten der Grossforaminifere *Heterostegina depressa* sind Diatomeen. *Naturwissenschaften* **63**, 486–7.
– (1978). The ultrastructure and taxonomic identity of the symbiotic algae of *Heterostegina depressa* (Foraminifera, Nummulitidae). *J. Mar. Bio. Asso. U. K.* **58**, 227–37.
Schoenberg, D. (1980a). An ecological view of specificity in algal-invertebrate associations, with reference to the associations of *Symbiodinium microadriaticum* and coelenterates. *In: Endocytobiology*, vol. 1, ed. W. Schwemmler & H. E. A. Schenk, pp. 146–54. Berlin: Walter de Gruyter & Co.
– (1980b). Intraspecific variation in a zooxanthella. *In: Endocytobiology*, vol. 1, ed. W. Schwemmler & H. E. A. Schenk, pp. 155–62. Berlin: Walter de Gruyter & Co.
Schoenberg, D., & Trench, R. (1980a). Genetic variation in *Symbiodinium* (=*Gymnodinium*) *microadriaticum* Freudenthal, and specificity in its symbiosis with marine invertebrates. I. Isoenzyme and soluble protein patterns of axenic cultures of *Symbiodinium microadriaticum. Proc. Royal Soc. Lond.* Series B. **207**, 405–27.
– (1980b). Genetic variation in *Symbiodinium* (=*Gymnodinium*) *microadriaticum* Freudenthal, and specificity in its symbiosis with marine invertebrates. II. Morphological variation in *Symbiodinium microadriaticum. Proc. Royal Soc. Lond.* Series B. **204**, 429–44.
– (1980c). Genetic variation in *Symbiodinium* (=*Gymnodinium*) *microadriaticum* Freudenthal, and specificity in its symbiosis with marine invertebrates. III. Specificity and infectivity of *Symbiodinium microadriaticum. Proc. Royal Soc. Lond.* Series B. **207**, 445–60.
Smith, D. C. (1975). Symbiosis and the biology of lichenised fungi. *In: Symbiosis*, vol. 39, ed. D. H. Jennings and D. L. Lee, pp. 373–405. Symposia of the Society for Experimental Biology. Cambridge: Cambridge University Press.
Smith, D., & Wiebe, W. (1977). Rates of carbon fixation, organic carbon release

and translocation in a reef-building foraminifer. *Maringopora vertebralis. Aust. J. Mar. Freshwater Res.* **28,** 311–19.

Sournia, A. (1976). Primary production of sands in the lagoon of an atoll and the role of foraminiferan symbionts. *Mar. Bio.* **37,** 29–32.

– (1977). Notes on primary productivity of coastal waters in the Gulf of Elat (Red Sea). *Internationale Revue der Gesamten Hydrobiologie* **62,** 813–19.

Taylor, D. (1974). Symbiotic marine algae: Taxonomy and biological fitness. *In: Symbiosis in the Sea,* ed. W. B. Vernberg, pp. 245–62. Columbia: University of South Carolina Press.

Trench, R. (1980). Integrative mechanisms in mutualistic symbioses. *Fifth Annual Colloquium.* Columbus: Ohio State University Press.

Weis, D. (1976). Digestion of added homologous algae by *Chlorella*-bearing *Paramecium bursaria. J. Protozoo.* **23,** 527–9.

Weter, G., Killingley, J., & Lutze, G. (in press). Stable isotopes in recent larger foraminifera. *Paleogeography, Paleoclimatology and Paleoecology.*

Winter, F. (1907). Zur Kenntnis der Thalamophoren. *In*: Untersuchung über *Peneropolis pertusus* (Forskat). *Archiv für Protistenk.* **10,** 1–113.

Woodruff, F., Saven, S., & Douglas, R. (1980). Biological fractination of oxygen and carbon isotopes by recent benthic foraminifera. *Mar. Micropaleontology* **5,** 3–11.

Zmiri, A., Kahan, D., Hochstein, S., & Reiss, Z. (1974). Phototaxis and thermotaxis in some species of *Amphistegina* (Foraminifera). *J. Protozoo.* **21,** 133–8.

Zucker, W. (1973). "Fine structure of planktonic foraminifera and their endosymbiotic algae." Ph.D. thesis, City University of New York.

4

The radiolarian symbiosis

O. ROGER ANDERSON
Biological Oceanography
Lamont-Doherty Geological Observatory of Columbia University
Palisades, NY 10964

The close association of cells with markedly different physiological characteristics and possibly distant phylogenetic affinities as occurs in plant-animal symbiosis raises the intriguing question of how these remarkable assemblages are initiated and sustained. It is clear that the close interaction of organisms with diverse metabolic and morphological properties places some constraint on (1) the kind of structural and metabolic union that can be established, (2) the degree to which mutual compatibility, and perhaps enhanced survival, is realized, and (3) the magnitude of the mutual gain to be achieved from the association.

A conceptual model of host-symbiont interaction

The magnitude of mutual gain, whether it be enhanced nutrition, greater protection, or improved environment for growth and reproduction, may be directly related to the diversity of the two potentially symbiotic organisms. The more diverse the metabolic and physiological qualities of the host and symbiont, the more likely it is that their combined activity may yield a markedly beneficial result for both members of the union. Thus, the strengths of one may supplement weaknesses in the other or, more dramatically, the combined activities of the two members may yield a much more novel metabolic or structural adaptation than could be predicted from the mere additive quality of their individual contributions. As a hypothetical example, each member may possess partial, but overlapping, sequences of an incomplete biochemical pathway that becomes fulfilled when the two organisms are united. Other cases include the marked mutual enhancement produced by nutrient exchange as occurs in some animal and algal associations. The two members are different

This work was supported in part by a grant from the Biological Oceanography Section of the National Science Foundation, No. OCE 80-05131. This is Bermuda Biological Station Contribution No. 881, Bellairs Research Institute Contribution No. 287, and Lamont-Doherty Geological Observatory Contribution No. 3442.

in trophic function, morphology, and subcellular structure, but their effective union may provide, among other benefits, protection of the symbiont and enhanced survival of the host through decreased dependence on the environment for sources of nourishment. The animal host may obtain nourishment from the primary productivity of the plant (e.g., Muscatine & Hand, 1958; Trench, 1971; Chalker & Taylor, 1975; Duguay & Taylor, 1978) while the plant may achieve protection against predation and perhaps utilize host excretory products as a primary nutrient source (Thorington & Margulis, 1981; see Chapter 3 in this volume). There is also evidence that some dinoflagellate symbionts possess sterols (Kokke, personal communication) and may secrete them along with other compounds into the environment, thus rendering the animal host less susceptible to predation or more resistant to invading disease organisms. Further examples of cooperative benefits to algal symbionts and their hosts are cited by Pardy (Chapters 1 & 5), Lee and McEnery (Chapter 3), and Ahmadjian and Jacobs (Chapter 8) in their contributions to this book. On the negative side, widely diverse organisms may incur varying degrees of mutual or unilateral incompatibility, causing exclusion of one member from the association or its destruction. Exclusion of potentially compatible members in a symbiotic association may be due to (1) structural incompatibilities arising from morphological features that prevent mutual accommodation (e.g., size constraints, inhibiting surface structures, or thickened boundaries preventing adequate interaction), or (2) physiological barriers (e.g., toxic excretory products, nutrient requirements precluded by association, or life cycle characteristics that prevent sustained association).

Hypothetically, organisms with diverse qualities may incur a decidedly large exclusion threshold when approaching a close association. An exclusion threshold is the combined set of factors that select against the association of two or more organisms. The higher the exclusion threshold, the less likely it is that potentially compatible organisms will associate. If, however, a high exclusion threshold is partially or wholly reduced by sufficient modification of the organisms to permit a closer union and/or by the magnitude of the mutual gain achieved from the union, a symbiotic association may nonetheless be established. In other words, factors contributing to incompatibility can be compensated for by (1) structural or physiological modifications in the potentially associative organisms that permit their symbiotic union, or (2) a mutual benefit so large that the exclusion threshold effect is reduced (i.e., the benefits outweigh the costs).

The three factors of mutual gain (G), exclusion effects (E), and adaptive modification (M) can be assembled in an interaction model to examine their effects on associative stability (S) of the potentially symbiotic organisms. The degree of associative stability is directly related to the mutual gain and is enhanced to the extent that adaptive modifications compensate for exclusion effects. Moreover, associations that have only a small mutual gain may achieve enhanced stability if there is sufficient adaptive modification to make their association more compatible. It is important to recognize that an adaptive modification need not result from a direct challenge by a potentially symbiotic organism. A phylogenetic change that may result from other selection pressures increases the probability of mutual compatibility. For example, a potential host may gradually become better adapted to its planktonic existence by an increased volume of bubblelike flotation chambers. This modification in response to selection pressure other than the symbiotic association may also increase the holding capacity for symbionts and favor a more stable association between host and symbiont. The conceptual model can be presented in a convenient shorthand notation as follows: $S = [G - f(E/M)]$, where $f(E/M)$ represents the extent to which adaptive modification compensates for exclusion effects. The function, as for example log (E/M), must be of the kind where $f(E/M)$ is positive in value when $E > M$, equals zero when $E = M$, and becomes negative when $E < M$, thus contributing to positive S.

Several interaction strategies can be mapped into the conceptual model. Symbiotic associations require that $G > 0$; there must be some mutual gain for the association to be classified as symbiotic. We can permit G to vary from very small values to appreciably large values. When G is small, and E is not very large, a small adaptive modification may be sufficient to compensate for the exclusion effect and yield a positive value for associative stability. A much larger adaptive modification ($M \gg E$) will increase the value of S, hence contributing to the overall stability of the association even though G is relatively small. When G is large, a decided tendency toward symbiosis is produced, provided the E/M balance is favorable. It is possible for associative stability to be enhanced, even though E is relatively large, by sufficient adaptive modifications in structure or function ($M > E$).

When $G \simeq 0$, and $E \simeq M$, or $E < M$, commensalism may occur, yielding a stable association but without substantial mutual gain; when, however, $G = 0$, and $E \geq M$, then exclusion or parasitism by one member of the other is indicated.

Our present knowledge of the quantitative relationships among the variables is too limited to permit a more refined development of the model. However, it may serve as a heuristic device to guide research toward a more systematic assessment of the factors in symbiotic associations and will be used in this chapter as a model to integrate current knowledge about algal associations with radiolarian hosts.

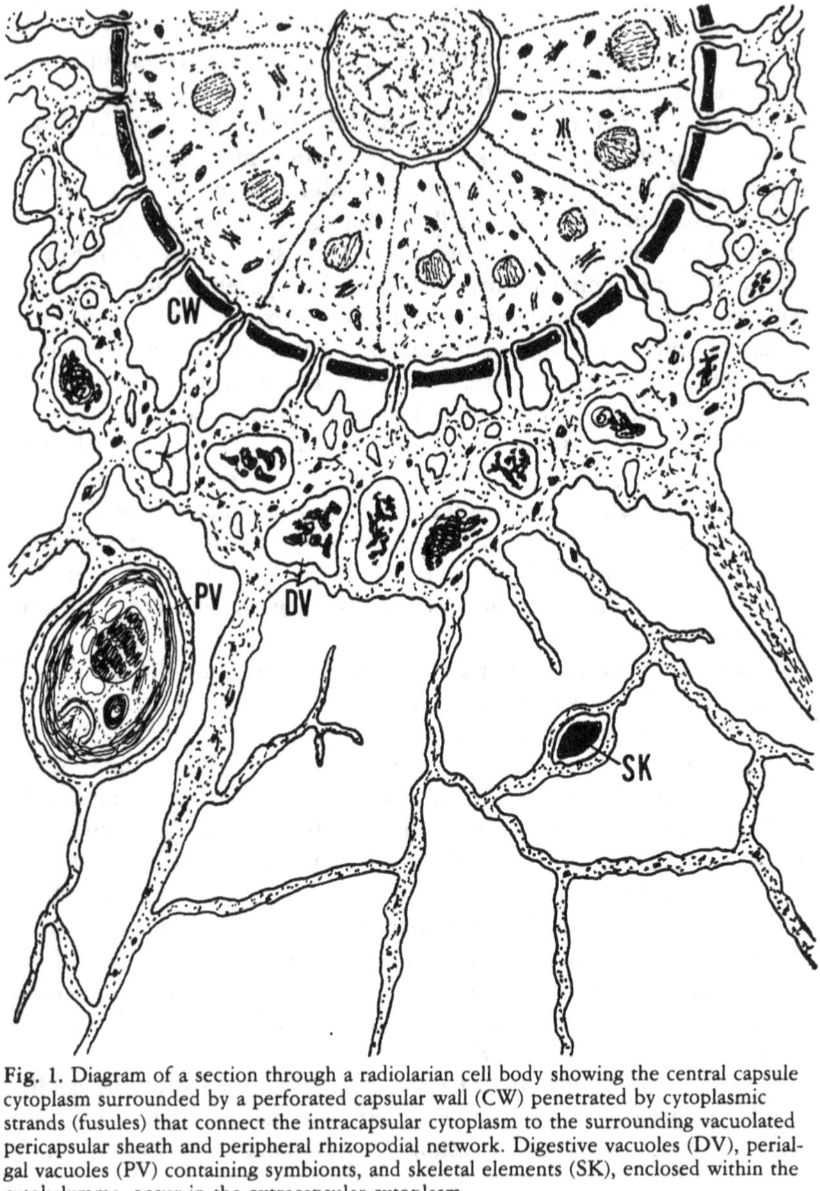

Fig. 1. Diagram of a section through a radiolarian cell body showing the central capsule cytoplasm surrounded by a perforated capsular wall (CW) penetrated by cytoplasmic strands (fusules) that connect the intracapsular cytoplasm to the surrounding vacuolated pericapsular sheath and peripheral rhizopodial network. Digestive vacuoles (DV), perialgal vacuoles (PV) containing symbionts, and skeletal elements (SK), enclosed within the cytokalymma, occur in the extracapsular cytoplasm.

Radiolarian morphology: the structure of the symbiont environment

In all symbiotic radiolarian species examined thus far, algal symbionts are held within a rhizopodial network surrounding the central cell body

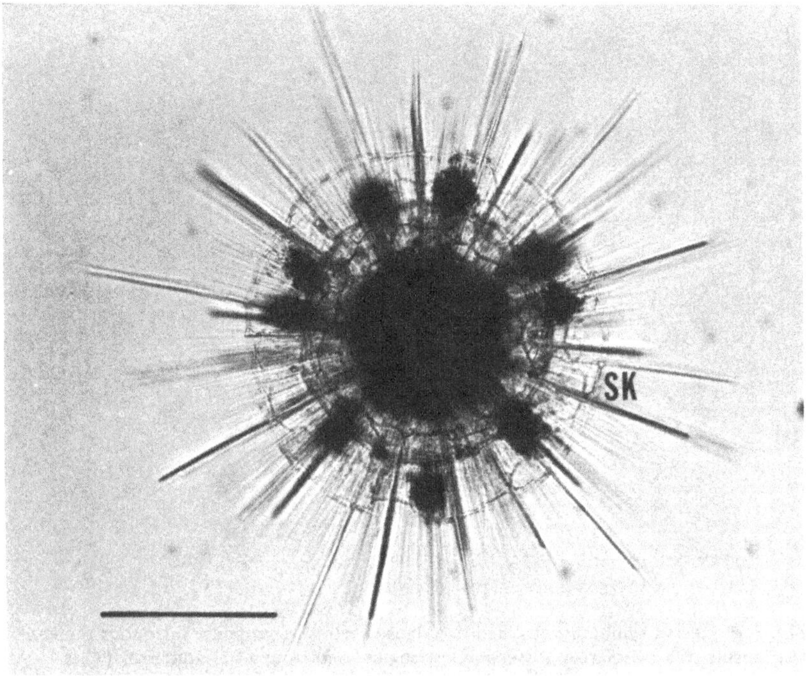

Fig. 2. A living solitary radiolarian. The opaque central capsule is surrounded by a halo of radially arranged axopodia penetrating a series of concentric lattice shells with long spines. SK, skeletal elements. Marker = 100 μm.

(Fig. 1), which contains one or more nuclei, food reserve bodies, mitochondria, and other vital organelles (Cachon & Cachon, 1972a, b; Anderson, 1976a, b, c; 1978b). The central cell body or central capsule is surrounded by a perforated organic capsular wall penetrated by protoplasmic strands that connect to the extracapsular cytoplasmic network. The extracapsular cytoplasm may form a vacuolated envelope surrounding the capsular wall and give rise to an elaborate network of rhizopodia, stiffened raylike axopodia and, where present, bubblelike alveoli that radiate outward around the central capsule. The proximal extracapsular cytoplasmic envelope is often rich in digestive vacuoles, compared to the more distal components of the extracapsulum, and it is presumed that digested food products pass directly into the central capsule through the protoplasmic strands in the capsular wall.

Various adaptive modifications in the extracapsular cytoplasm provide increased surface area and complexity of the rhizopodial network. In some species axopodia, consisting of stiffened pencillike strands of cytoplasm containing a central shaft of microtubules, radiate out from the central capsule, thereby increasing the total surface area of the cytoplasm

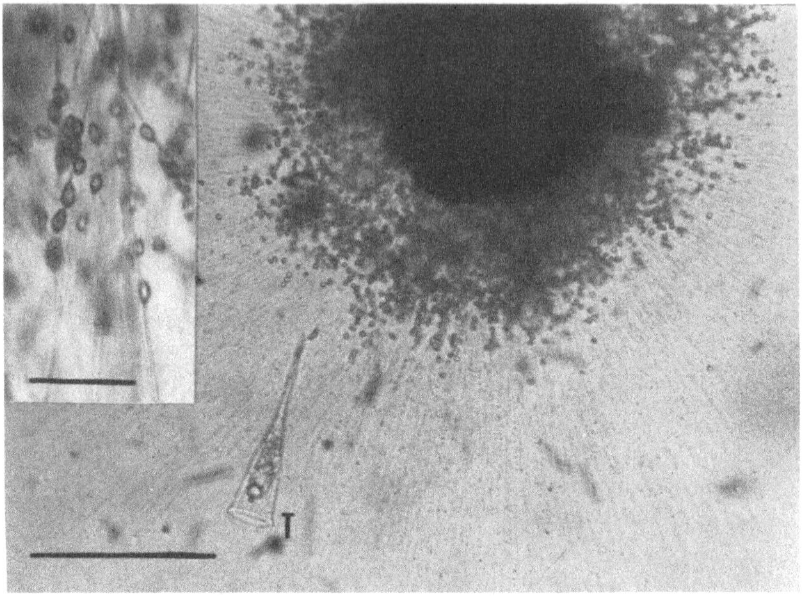

Fig. 3. A halo of symbionts attached to a dense corona of axopodia surrounds the central capsule of a radiolarian possessing a spongiose skeleton; a tintinnid prey (T) is snared in the rhizopodia. Marker = 400 μm. *Inset*: Detail of symbionts attached to rhizopodia. Marker = 100 μm.

and possibly providing mechanical support for the surrounding web of rhizopodia. A skeleton composed of amorphous silica is characteristic of many radiolarian species and often possesses an intricate design consisting of perforated and diversely ornamented hollow spheres bearing spines of varying length. In some species, the spines may be much longer than the diameter of the shell. In other species, the skeleton consists of a conical or spherical shell composed of a geodesiclike lattice. Rhizopodia stream along the surface of the skeleton, or form a sheath around it, and use these anchor points to attach an elaborate network of cytoplasmic strands that further branch near the periphery of the cell, yielding long, raylike cytoplasmic filaments extending outward into the surrounding environment (Figs. 2, 3). The elaborate perforations in the skeleton and the long spines, when present, provide a large surface area that may facilitate prey capture and enhance the symbiont-holding capacity of the extracapsular network. Some of the larger skeletonless species, such as *Thalassicolla nucleata* Huxley (with diameters up to 3 mm), possess a massive array of bubblelike alveoli that increase the cytoplasmic area for rhizopodial and symbiont attachment (Fig. 4).

In colonial radiolarian species, possessing hundreds of cells within a

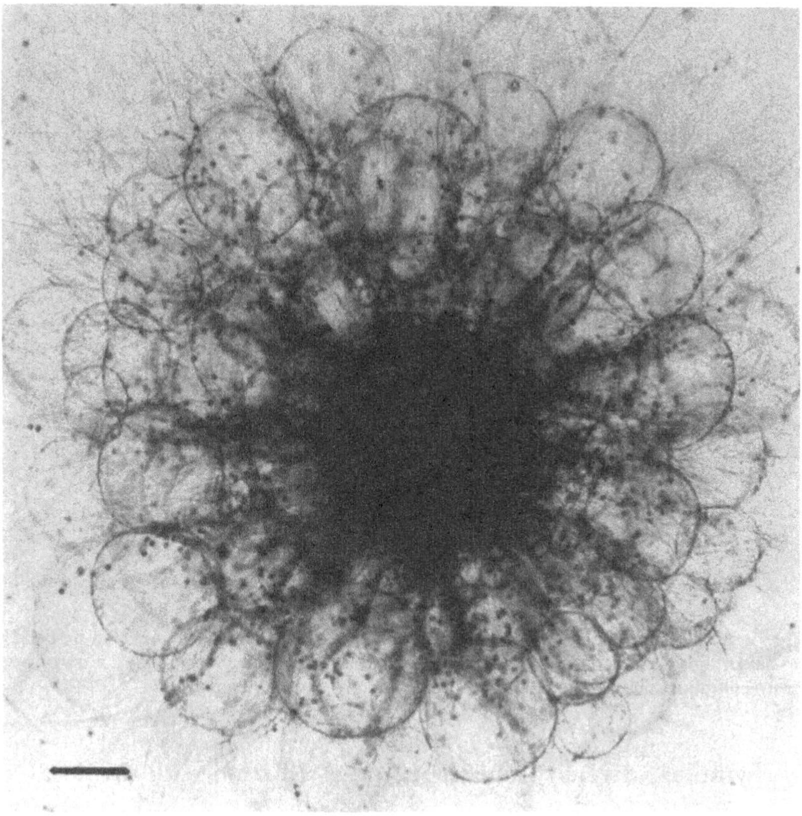

Fig. 4. *T. nucleata.* A mass of bubblelike alveoli, penetrated by rhizopodial strands, surrounds the central capsule. Marker = 300 μm. Reproduced by permission from Anderson (1978b)

translucent gelatinous envelope, the rhizopodial network and peripheral corona of radiating cytoplasmic filaments are substantially developed (Fig. 5). The colonies may be extremely large by standards of protozoan dimensions, reaching diameters of several centimeters for spherical forms, or lengths up to 1 m for filiform colonies. The increase in surface area provided by the elaborate interconnected rhizopodial network among cells in the colony and the enlarged surface area created by the supporting gelatinous sheath favor an optimum distribution of algal symbionts for light gathering and, possibly, nutrient exchange. Moreover, the gelatinous translucent wall of the colony provides a clear advantage for maximizing symbiont primary productivity while simultaneously maintaining control over symbiont distribution.

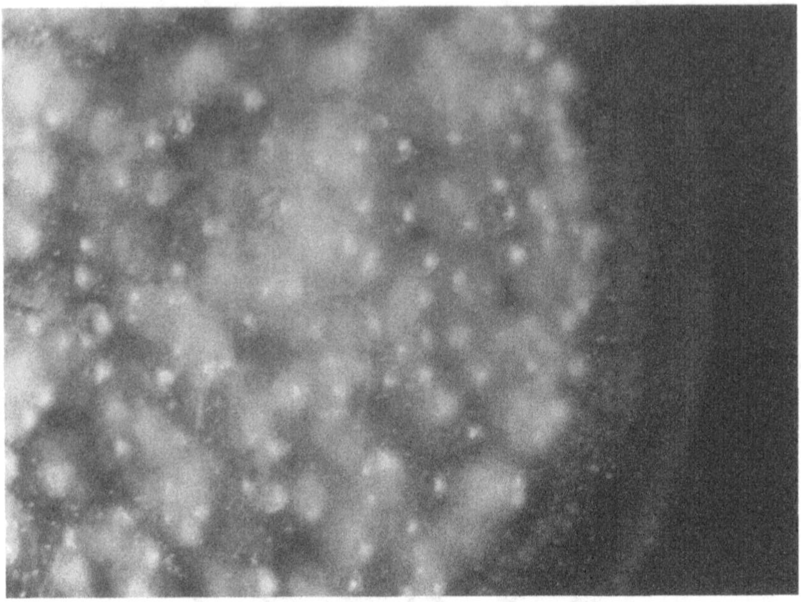

Fig. 5. A colonial radiolarian possesses hundreds of cell bodies interconnected by rhizopodial strands containing numerous dinoflagellate symbionts enclosed within a translucent gelatinous sheath. The colony is approximately 2 cm in diameter.

Symbiont diversity and structural association with the host

Dinoflagellates are among the most commonly observed algal symbionts in many colonial and some solitary radiolaria (Fig. 6). At present, we do not know the species of the symbionts. None have been isolated or cultured in the laboratory, as is required to make taxonomic identifications based on thecal morphology. They exhibit typical dinoflagellate fine structure, including a prominent mesokaryotic nucleus with puffy, coiled chromosomes and a peripheral system of membranous cisternae surrounding the cell (Anderson, 1976a). A prasinophyte symbiont (Fig. 7) has been observed in some solitary species possessing spongiose skeletons and belonging to the family Astrosphaeridae. The symbionts possess a typical prasinophyte cellular organization, including a nuclear invagination into the pyrenoid and multiple flagella with surface scales (Anderson, 1976a). They resemble the prasinophyte symbiont *Platymonas convolutae* found in the marine flatworm *Convoluta roscoffensis* (Graff) examined by Parke and Manton (1967). Intracellular algae have also been reported in acantharia, a group of protozoans related to radiolaria but possessing strontium sulfate skeletons. One group of symbionts are Dinophyceae containing trichocysts in the cytoplasm, and others are of uncertain taxonomic position, characterized, among other

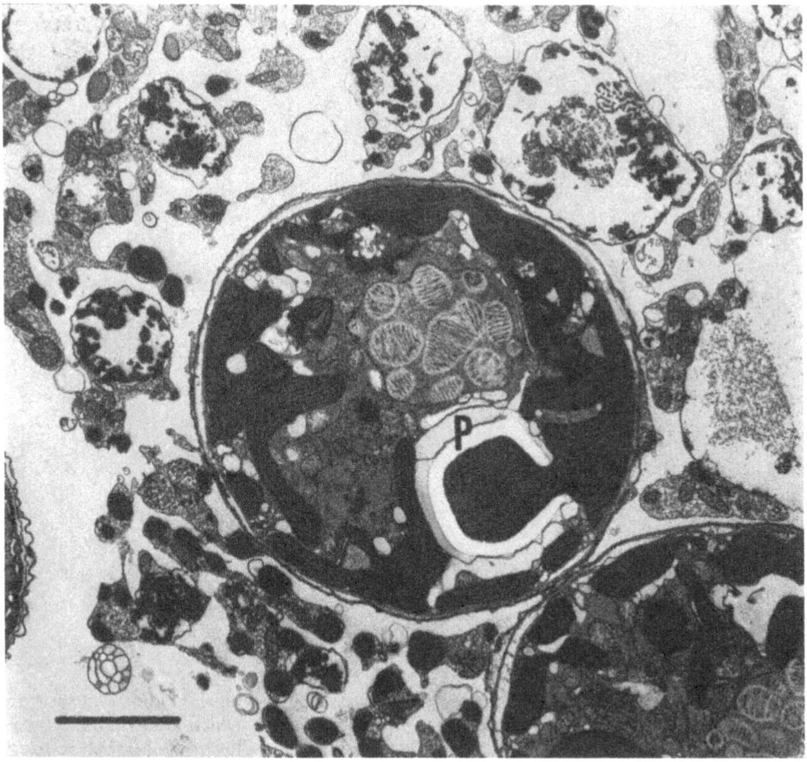

Fig. 6. Fine structure of a dinoflagellate symbiont enclosed within a vacuole in the radiolarian rhizopodial network. The nucleus contains puffy, coiled chromosomes. Chloroplast lobes occur at the periphery of the cell, and the pyrenoid (P) has a deposit of starch around it. Marker = 5 μm.

things, by numerous discoidal plastids, each containing an interlamellar pyrenoid (Hollande & Carre, 1974). More recently, Febvre and Febvre-Chevalier (1979) have presented evidence that at least one group of large, brown-pigmented algal symbionts in acantharia belongs to the order Prymnesiales.

Our knowledge of radiolarian symbiont diversity is limited by the relatively few modern studies on the fine structure of radiolarian symbionts. A clear advantage will be attained if the symbionts can be isolated in culture and their morphological and nutritional characteristics more firmly established.

Current fine structure evidence provides insight into the structural association of symbiont with host and adaptive modifications that accompany the symbiotic state. All of the dinoflagellate symbionts observed in radiolaria are coccoid, that is, they have become somewhat more rounded and have shed their flagella with exception of the flagellar

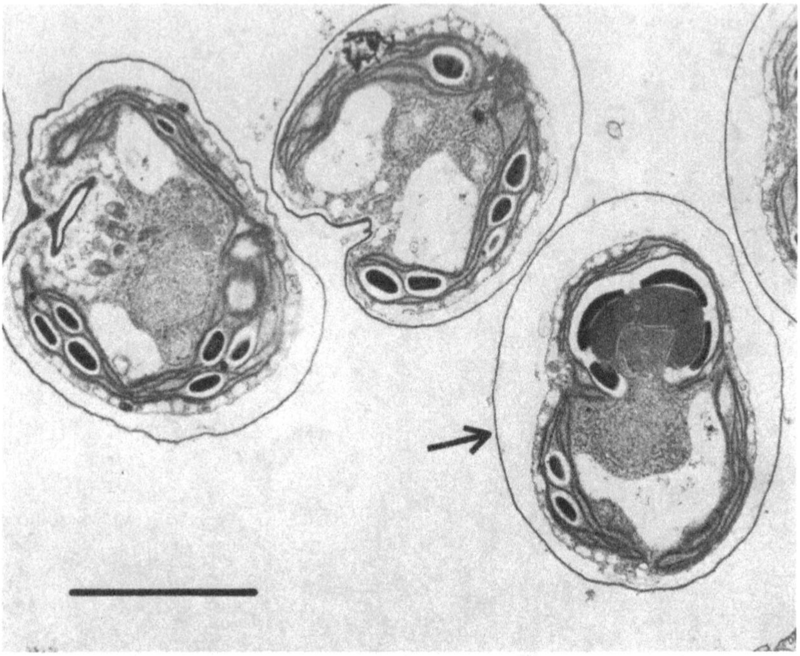

Fig. 7. Fine structure of a prasinophyte symbiont associated with the radiolarian shown in Fig. 3. A lobe of the nucleus invaginates into the pyrenoid, which is surrounded by a starch sheath. The cell is surrounded by a thin osmiophilic wall (arrow). Marker = μm.

bases, which are observable in the fine structure views of ultrathin sections. The thecal plates that compose the wall of the dinoflagellate are lost, and only a very thin organic membrane is present in the peripheral vesicles that normally contain the plates. The loss of flagella clearly brings the immobile dinoflagellate under greater control of the host, and the absence of a thecal wall permits a close structural and perhaps physiological association between host and symbiont. Dinoflagellate symbionts are usually enclosed within a perialgal vacuole located in the rhizopodial network or near the vacuolated layer surrounding the capsular wall. Enclosure of the dinoflagellate within a vacuole permits constant control over the symbiont by the host and may permit regulated exchange of materials across the vacuolar membrane. Clear evidence of host control over symbiont distribution is shown by the rhizopodial and axopodial streaming activity whereby the position of the symbionts may be changed dramatically from deep within the proximal layer of extracapsular cytoplasm to a very distal position near the rhizopodial periphery. Indeed, a regular diurnal cycle of symbiont distribution occurs: The symbionts are withdrawn at night in close proximity to the central capsule wall (Fig. 8), but at the onset of daylight, most of the symbionts are distributed

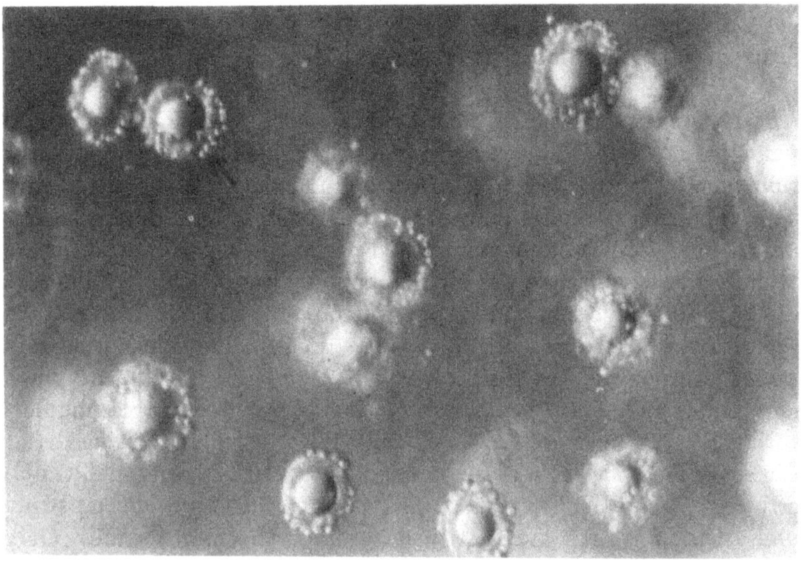

Fig. 8. Colonial radiolarian central capsules with dinoflagellate symbionts (arrow) drawn close to the pericapsular layer as they appear at night. Each central capsule is approximately 100 μm in diameter.

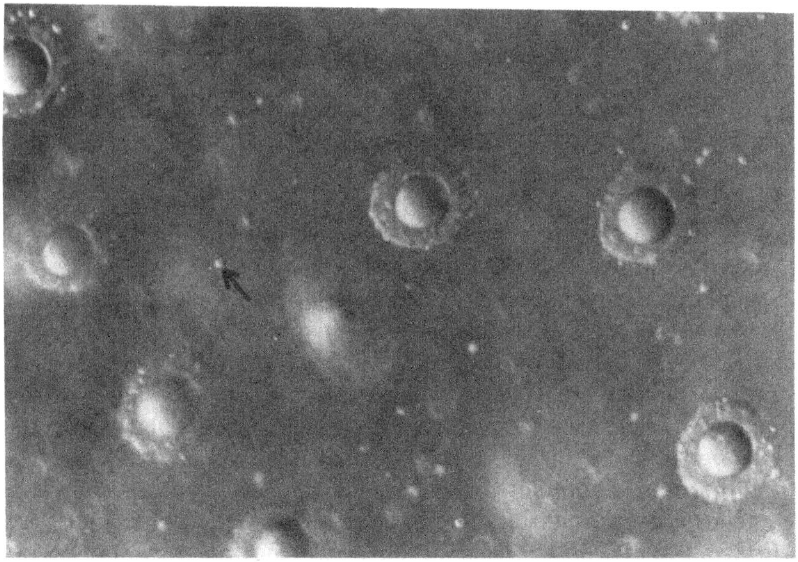

Fig. 9. A view of the colony shown in Fig. 8, during the daylight phase of the cycle when the symbionts (arrow) are distributed in the rhizopodial network.

into the peripheral cytoplasm (Fig. 9). This periodic pattern has been observed in colonial and solitary species.

A marked degree of host cytoplasmic differentiation occurs in the pericapsular cytoplasmic sheath. Perialgal vacuoles containing healthy symbionts are often observed in close proximity to digestive vacuoles containing remains of ingested prey, thus indicating a subtle mechanism of vacuolar differentiation to mediate protection of symbionts while simultaneously consuming ingested food particles. At a later point in the chapter, the role of symbionts as potential prey for the host is discussed.

The radiolaria-bearing prasinophyte symbionts may not have as close a structural association with the algae as do the dinoflagellate-bearing species. Among the few cases observed (Anderson, 1976a), the prasinophyte symbionts are held more loosely within the cytoplasmic network. The symbionts are "coccoid" and have only stubby flagellar bases bearing scales. The cell is surrounded by a distinct osmiophilic wall (arrow, Fig. 7), which suggests on the whole a much more distant structural association with the host than observed in dinoflagellates. Only occasional surface contact between the host rhizopodia and symbiont wall is observed. The large number of algae associated with the host (Fig. 3) and their clear attachment to the rhizopodial surface strengthen the assumption that they are symbionts. The presence of prasinophyte symbionts, loosely held in the extracapsular cytoplasm, has also been reported by Cachon and Caram (1979) in the large gelatinous radiolarian *Thalassolampe margarodes*.

Host-symbiont interactions

Much of our knowledge about host-symbiont interactions in radiolaria prior to recent years (e.g., Anderson 1976a, 1978a, 1980, 1983) comes from research in the late nineteenth and early twentieth centuries (Brandt, 1883, 1885; Gamble, 1909). Many earlier studies suffered from obvious shortcomings due to the lack of sophisticated techniques needed to trace physiological events. Nonetheless, these pioneering studies established that the symbionts provide nourishment for the host as shown by enhanced survival of symbiont-bearing radiolaria grown in food-free cultures in the light compared to similar cultures placed in darkness. However, neither the kind of nourishment nor the mechanism of its uptake by the host was determined. Moreover, a prevailing, and erroneous, opinion was that the main supportive role of the symbiont was to supply oxygen for the animal host. Brandt (1882) observed starch grains in the symbiotic algae and suggested that carbohydrates may be transferred to the host. The question of whether the symbionts were ingested from time to time as food or only secreted food substances absorbed by the host was apparently a topic of heated debate. Recent

Table 1. *Comparative data for symbiont primary productivity and some free-living algae*

Alga	Cell diameter (μm)	Primary productivity μg C/cell/min
Dunaliella primolecta	5–7	44×10^{-9}
Collozoum longiforme symbiont	15	160×10^{-9}
Collosphaera globularis symbiont	15–20	350×10^{-9}
Gonyaulax polyedra	30–50	301×10^{-8}

advances in radioisotope labeling and electron microscopic techniques have helped to elucidate some of these questions, although our knowledge remains limited.

Symbiont primary productivity

The rate of symbiont primary productivity within the host has been determined using ^{14}C isotope labeling techniques. Table 1 presents comparative data for radiolarian symbionts and some free-living algae in relation to their size (Anderson, 1980). The rate of carbon fixation in symbiotic dinoflagellates is comparable to free-living cells. Moreover, electron microscopic investigations of healthy radiolaria show that the symbionts contain abundant starch and food reserves, suggesting they are not under trophic stress.

It is not known to what extent the radiolarian may control symbiont primary productivity, their release of photosynthetically derived substances, or their rate of reproduction. Some recent research by Swanberg and Harbison (1981) suggests that the amount of food ingested by some colonial species may influence the rate of symbiont primary productivity. Symbionts in hosts with large amounts of prey showed higher rates of primary productivity than those in hosts with less prey. These preliminary data raise interesting questions about the mechanism of this regulation. Is it mediated by host control of the symbiont through chemical or physical hormonal-like regulators, or is it an indirect effect of excretory products released by the host that serve as nutrients for the symbionts and influence their rate of metabolism? Additional research will be needed to clarify this interactive effect.

Symbiont-derived nutrition

Evidence for translocation of photosynthetically derived carbon compounds from symbiont to host has been obtained with the colonial radiolarian *Collosphaera globularis* Haeckel (Anderson, 1978a).

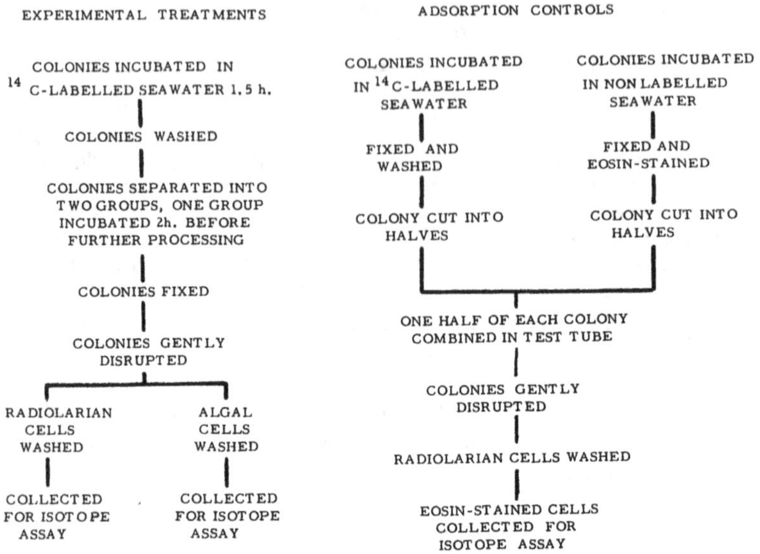

Fig. 10. Flow diagram of a controlled experiment to determine if photosynthetically derived substances are translocated from symbiont to host (Anderson, 1978a).

This species is particularly amenable to radioisotopic analysis of symbiont-host interactions since the radiolarian central capsule is surrounded by a microperforated shell that prevents the symbionts from passing into the pericapsular cytoplasm. The host cells can be carefully and totally separated from the symbionts by gentle disruption of the colony, repeated washing of the host cells to remove symbionts, examination for symbiont contamination using light optics, and collection of the cleaned host cells using a micropipette. Intact colonies may be incubated in seawater containing $H\,^{14}CO_3^-$ to assess symbiont primary productivity in situ and to determine if ^{14}C-labeled photosynthetically derived substances are translocated from algae to host. An experimental plan (Anderson, 1978a, 1980) is shown in Fig. 10.

In addition to the colonies incubated in the light with ^{14}C-labeled seawater, a dark control in identical culture medium was used to determine how much ^{14}C uptake occurs in the absence of light (nonphotosynthetic incorporation). Moreover, two adsorption control preparations were used. An intact glutaraldehyde-killed colony was incubated in the ^{14}C-labeled culture medium to determine surface adsorption. To control for possible spurious transfer of ^{14}C-labeled compounds from symbiont to host during colony disruption and isolation of host cells, an eosin-stained colony, free of ^{14}C label, was gently disrupted in the standard way in the presence of a colony incubated in ^{14}C-labeled seawater. Only the eosin-stained cells were isolated with a micropipette and the amount

Table 2. *^{14}C Isotopic evidence for host assimilation of symbiont-derived organic compounds in* C. globularis

Sample	Incubation (h)	Radioactivity minus blank (dpm)
Host cells (100)	1.5	21.7
Host cells (100)	3.5	51.6
Symbionts (300)	1.5	447.0
Dark control cells (100)	1.5	1.3
Adsorption control cells	—	<1.0

Source: From Anderson (1978a).

of contamination assessed using a standard liquid scintillation counting technique. To gain further evidence of translocation of organic compounds, some light-incubated colonies were washed free of ^{14}C-labeled medium and incubated for an additional 2 h to determine if there were additional accumulation of isotopic-labeled compounds by the host.

The results of the experiment are presented in Table 2. The host cells isolated from the light-incubated colonies show significant incorporation of ^{14}C-labeled compounds compared to the control groups. Moreover, the colonies placed in seawater for an additional 2 h following incubation in ^{14}C-labeled medium showed an increase in accumulation of ^{14}C-labeled organic compounds, thus providing additional evidence of translocation. Further research with large quantities of cells will be required to determine the rate of translocation and total host accumulation in a given period of time.

The nutritional gain achieved by the host through assimilation of symbiont-derived organic substances may contribute to the stability of the host-symbiont association. It remains to be determined what nutritional benefit, if any, the symbiont may realize.

Host ingestion of algal symbionts

The degree of interdependence and mutual gain versus unilateral benefit realized by members of a symbiotic union may vary considerably among species. In some cases, the symbionts may not be entirely free of exploitation or exclusion by the host. Indeed, this is the significance in part of the (*E*/*M*) function in the conceptual model. Among the strategies for interaction adopted by symbiotic organisms we may expect to find a range of values for host-symbiont modifications that offset tendencies toward exploitation of one member by the other. In *Paramecium bursaria* Pringsheim, infection of aposymbiotic hosts by potentially symbiotic

Chlorella sp. depends upon the relative effectiveness of the algae in inhibiting host digestive processes (Karakashian, 1975). The early stages of host infection may be limited by host digestion of some of the algae, and it appears that a certain critical number of algae must be present to establish a stable association. When a stable association is achieved, and the host is supplied with an external food source, very few symbionts are digested. The degree of host susceptibility to algal infection may vary with different host strains, suggesting variability in the extent to which a symbiotic association provides unilateral gain for one member over the other. Similar adaptive modifications in algal-fungal associations are cited by Ahmadjian and Jacobs in Chapter 8 in this volume. A certain degree of unilateral gain at the expense of one member in a symbiotic association may be accepted if the mutual gain on the whole is sufficiently large. It is sometimes argued that if the symbionts are digested by the host this would preclude a priori a symbiotic relationship. This position is based apparently on the notion that each individual organism in the association must benefit if the union is to be classified as symbiotic. A more appropriate focus would be on the population of cells in the association rather than on individual organisms. If the animal host digests some of the algae, but on the whole sustains a vigorous, reproducing algal population, this should qualify as a symbiotic association. From a quantitative viewpoint, therefore, the problem becomes one of determining the degree of symbiotic association rather than presuming an all-or-none model.

There is evidence of host digestion of algal symbionts by the colonial radiolarian *Collozoum inerme* Müller (Anderson, 1976b). When the colony is illuminated in laboratory cultures using cool white fluorescent lights (154 $\mu W/cm^2$ in the range of 400–500 nm and 77 $\mu W/cm^2$ in the range of 500–700 nm) the radiolarian contains a steady, vigorous population of dinoflagellate symbionts with about 20–30 algae per host cell. Electron microscopic observations show the algae are dividing and contain an appreciable reserve of carbohydrate in the pyrenoid cap. However, when the colonies are placed in the dark, the number of algae decreased by 10%–15% in the first 4 days. Light microscopic observations of the dark-incubated colonies reveal that many of the algal cells are sequestered into vacuoles in the pericapsular cytoplasmic layer. They are crenated and dark yellow in color, suggesting they are being digested. The remainder of the algae in the rhizopodial system appear normal. Control colonies *maintained in the light* show no decline in algal population density even though there is evidence of symbiont ingestion by the host (Fig. 11). These data imply that illuminated colonies may ingest algal symbionts, but at a rate that compensates for algal reproduction, thereby maintaining a nearly steady algal population.

Direct evidence of host digestion of symbiont cells was obtained by

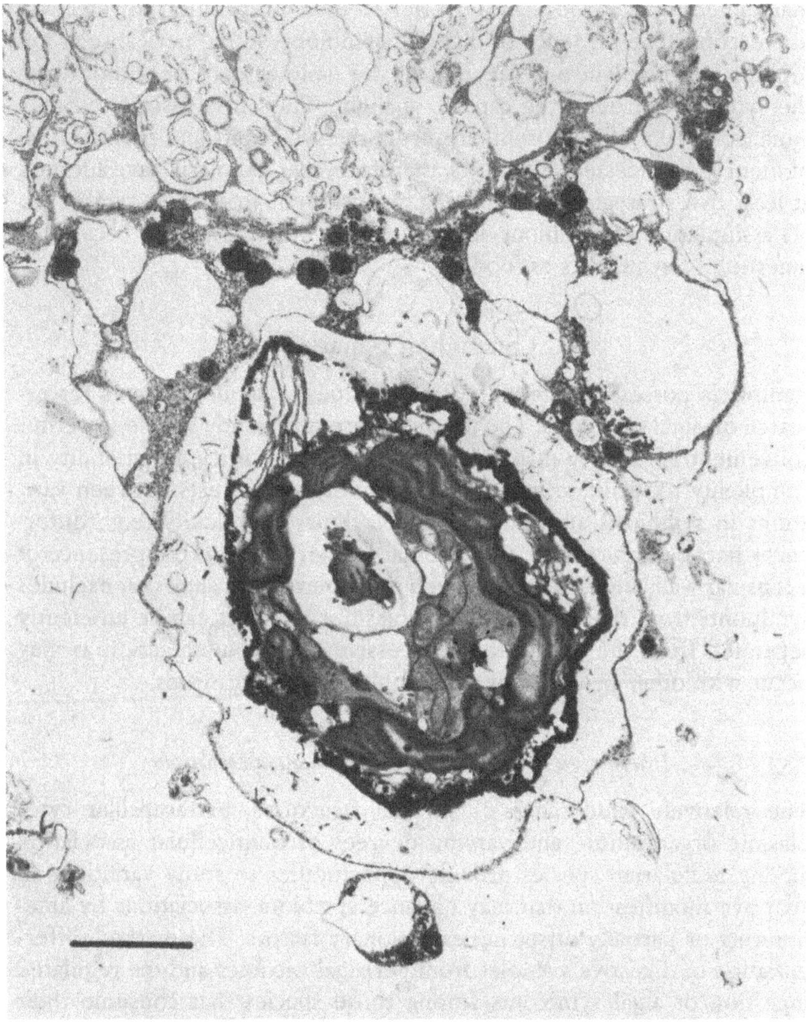

Fig. 11. Evidence for digestion of symbionts in *C. inerme*. The dinoflagellate symbiont is enclosed in a digestive vacuole containing a dense deposit of acid phosphatase reaction product. Marker = 2 μm.

electron microscopic cytochemical techniques. Illuminated colonies maintained in Millipore-filtered, algal-free seawater were prepared for electron microscopy using a histochemical stain for lysosomal acid phosphatase. Ultrathin sections of the colony (Fig. 11) exhibited dinoflagellate algae in digestive vacuoles containing dense deposits of acid phosphatase reaction product, thus confirming light microscopic observations indicating symbiont digestion. The dinoflagellate prey in the digestive vacuoles are undoubtedly symbionts since the cultures of the radiolaria were maintained in alga-free seawater prior to their preparation for electron microscopy. The presence of digested symbionts in the pericapsular cytoplasm may explain one function of the daily rhythm of withdrawing the symbionts around the capsule at night. This close association may enhance uptake of photosynthates from the symbionts and permit some of them to be ingested. It appears, therefore, that radiolaria have adopted at least two strategies for obtaining nourishment from their symbionts: (1) assimilation of symbiont-derived photosynthates, and (2) occasional ingestion of symbionts as food.

Concluding remarks

Radiolaria possess several appealing qualities as model systems for research on algal symbiosis. The organisms are sufficiently simple to permit convenient laboratory manipulation, yet exhibit sufficient variability in complexity to facilitate inquiry on the interactive effects between variations in symbiosis and the degree of cellular complexity (e.g., differences between unicellular and colonial species). Due to the presence of a capsular wall, and in some species a microperforated shell that excludes symbionts from the central cell cytoplasm, symbionts can be efficiently separated from the host without excessive cross-contamination as may occur with other protozoa or some multicellular organisms.

Interaction strategies in host-symbiont association

The relatively wide range of skeletal structures, extracapsular cytoplasmic organization, and varying degrees of multicellular association among radiolarian species provide opportunities to study variations in adaptive modification that may enhance symbiotic associations by ameliorating or partially offsetting exclusionary factors. The marked differentiation of digestive vacuoles from perialgal vacuoles and the regulated ingestion of algal symbionts among those species that consume their symbionts counterbalance an exclusionary tendency for the host to prey excessively on the symbiont population. Likewise, the adaptive modifications for increased surface area mediated by skeletal complexity, alveolar close packing that supports an extensive rhizopodial network, radially arranged axopodial systems, and, in colonial forms, a massive gelatinous envelope contribute toward enhanced symbiont-holding capacity under conditions that favor exchange with the environment and primary productivity. The degree of mutual gain for radiolarian and symbiont is only beginning to be elucidated. It is clear, however, that the host in several species is able to gain nutrition from the symbionts and that the symbionts photosynthesize at basic rates comparable to free-living algae, reproduce in situ, and may exhibit enhanced primary productivity through nutrient uptake released by the host during periods of elevated metabolism. The diversity of the symbionts (including prasinophytes and dinoflagellates) suggests that a certain degree of plasticity has occurred in the radiolarian solution to the problem of establishing a mutually beneficial association with genetically diverse organisms.

Symbiosis and the origin of primitive multicellular systems

The gelatinous, colonial radiolaria may provide insight into the biological processes mediating cellular association into primitive multicellular organisms. All of the colonial species examined possess symbiotic algae. The presence of symbionts in primitive aggregates in interconnected cells could reinforce adaptive trends toward closer coordination among the colony-building cells, their increased network complexity, and greater cell number. Aggregates of host cells linked in a loose rhizopodial network can capture and may retain larger numbers of symbionts than solitary cells. This is particularly likely when a gelatinous coat of large surface area is secreted around the cells, thus increasing the total area available for symbionts. Any tendency toward increased volume by the multicellular aggregate and greater cooperation among the cells is likely to enhance the rhizopodial surface area and hence the total carrying capacity of the system for the symbionts, thereby increasing total primary productivity and permitting greater control over symbiont distribution. The presence of symbionts may also reinforce trends toward multicellularity by providing a source of constant nourishment for the complex assembly under conditions of oligotrophic nutrition. Hypothetically, symbionts may have served an enabling role in sustaining primitive multicellular associations, particularly during the early period of colony development. Recent data from research in our laboratory have further demonstrated the significant enabling role of algal symbionts in providing nourishment for the radiolarian host. Symbiont-bearing specimens of the large solitary radiolarian *T. nucleata* receive a substantial proportion of their sustenance from the symbionts. Illuminated symbiont-bearing specimens maintained in laboratory culture, but not fed, lived on the average as long as comparable specimens fed *Artemia nauplii*.

The obscure phylogenetic origins of some metazoa may rest in part on the fortuitous association of algae and protozoa that provided a necessary trophic support for simple cells to aggregate into a fragile, but highly productive, primitive assembly and thus initiate the long phylogenetic procession from unicells to higher forms of life.

Summary

An analysis of host-symbiont interaction in radiolaria and other symbiont-bearing organisms has been incorporated in a conceptual scheme relating host-symbiont associative stability to the degree of mutual gain and the amount of their adaptive modification in relation to exclusion variables between host and symbiont. Radiolaria exhibit a wide range of structural characteristics favoring symbiont association, including an extensive extracapsular cytoplasmic network sometimes supported by an elaborate siliceous skeleton or a jellylike capsule composed of closely packed bubblelike alveoli that increase the surface area and enhance symbiont-holding capacity. Symbiotic algae include dinoflagellates and prasinophytes. Primary productivity of symbionts in situ is comparable to free-living algae when corrected for differences in size. A variety of structural adaptations by the host and alga have developed to stabilize their association, including host rhizopodial differentiation to distribute the algae for optimum light exposure and perhaps efficient exchange with the environment, differentiation of the host vacuolar system to protect symbionts against unregulated digestion, and enclosure of the symbiont in some species within perialgal vacuoles. The symbionts are modified by becoming nonmotile and by losing their wall, thus bringing them in closer contact with the host.

Radiolaria obtain nourishment from the symbionts by assimilation of symbiont-derived photosynthates or occasional ingestion of the symbionts. There is limited evidence that the symbionts may assimilate metabolic waste products from the host or otherwise depend on the host for regulation of their rate of primary productivity. The complex organization of symbiont-bearing colonial radiolaria may provide a model system to study the biological factors contributing to the origin of primitive multicellular systems.

References

Anderson, O. R. (1976a). A cytoplasmic fine-structure study of two spumellarian Radiolaria and their symbionts. *Mar. Micropaleontol.* **1**, 81–99.

– (1976b). Ultrastructure of a colonial radiolarian *Collozoum inerme* (Müller) and a cytochemical determination of the role of its Zooxanthellae. *Tissue and Cell* **8**, 195–208.

– (1976c). Fine structure of a collodarian radiolarian (*Sphaerozoum punctatum* Müller 1858) and cytoplasmic changes during reproduction. *Mar. Micropaleontol.* **1**, 287–97.

– (1978a). Fine structure of a symbiont-bearing colonial radiolarian, *Collosphaera globularis*, and ^{14}C isotopic evidence for assimilation of organic substances from its zooxanthellae. *J. Ultrastructure Res.* **62**, 181–9.

– (1978b). Light and electron microscopic observations of feeding behavior, nutrition, and reproduction in laboratory cultures of *Thalassicolla nucleata*. *Tissue and Cell* **10**, 401–12.

– (1980). Radiolaria. *In*: M. Levandowsky & S. H. Hutner, eds., *Biochemistry and Physiology of Protozoa*, 2nd ed., vol. 3, pp. 1–42. New York: Academic Press.

– (1981). Radiolarian fine structure. *In*: T. Simpson & B. E. Volcani, eds., *Silicon and Siliceous Structures in Biological Systems*, pp. 347–79. New York: Springer-Verlag.

– (1983). *Radiolaria*. New York: Springer-Verlag.

Brandt, K. (1882). Ueber die morphologische und physiologische Bedeutung des Chlorophylls bei Thieren. *Mittheil. Zool. Stat. Neapel* 4, 191–302.

– (1883). Ueber Symbiose von Algen und Thieren. *Separat Abzug aus Arch. f. Anat. u. Physiol. Phys. Abetil.* 445–54.

– (1885). Die koloniebildenden Radiolarien (Sphaerozoëen) des Golfes von Neapel und der angrenzenden Meeresabschnitte. *Monogr. Fauna und Flora Golfes Neapel* **13**, 1–276.

Cachon, J., & Cachon, M. (1972a). Le système axopodial des Radiolaires Sphaeroidés. I. Centroaxoplastidies. *Arch. Protistenk.* **114**, 51–64.

– (1972b). Le système axopodial des Radiolaires Sphaeroidés. II. Les périaxoplastidiés. III. Les crytoaxoplastidiés (anaxoplastidiés). IV. Les fusules et le système rhéoplasmique. *Arch. Protistenk.* **114**, 291–307.

Cachon, M., & Caram, B. (1979). A symbiotic green alga, *Pedinomonas symbiotica* sp. nov. (Prasinophyceae), in the radiolarian *Thalassolampe margarodes*. *Phycologia* **18**, 177–84.

Chalker, B. E., & Taylor, D. L. (1975). Light-enhanced calcification and the role of oxidative phosphorylation in calcification of the coral *Acropora cervicornis*. *Proc. Royal Soc. Lond.* Series B. **190**, 323–31.

Duguary, L. E., & Taylor, D. L. (1978). Primary production and calcification by the soritid Foraminifer *Archais angulatus* (Fichtel & Moll). *J. Protozool.* **25**, 356–61.

Febvre, J., & Febvre-Chevalier, C. (1979). Ultrastructural study of zooxanthellae of three species of acantharia (Protozoa: Actinopoda), with details of their taxonomic position in the Prymnesiales (Prymnesiophyceae, Hibberd, 1976). *J. Mar. Biol. Ass. U. K.* **59**, 215–26.

Gamble, F. W. (1909). The protozoa (Section E – The Radiolaria). *In*: E. R. Lankester, ed., *A Treatise on Zoology*, pt. 1, p. 117. London: Adam and Charles Black.

Hollande, A., & Carre, D. (1974). Les Xanthelles des Radiolaires Sphaerocollides des Acanthaires et de *Velella velella*: infrastructure cytochimie, Taxonomie. *Protistologia* **10**, 573–602.

Karakashian, M. W. (1975). Symbiosis in *Paramecium bursaria*. *In: Symposia of the Society for Experimental Biology*, no. 29, ed. D. H. Jennings & D. L. Lee, pp. 145–73. Cambridge: Cambridge University Press.

Muscatine, L., & Hand, C. (1958). Direct evidence for the transfer of materials from symbiotic algae to the tissues of a coelenterate. *Proc. Nat. Acad. Sci.* **44**, 1259–63.

Parke, M., & Manton, I. (1967). The specific identity of the algal symbiont in *Convoluta roscoffensis*. *J. Mar. Biol. Ass. U. K.* 47, 445–64.

Swanberg, N. R., & Harbison, G. R. (1981). The ecology of *Collozoum longiforme*, sp. nov., a new colonial radiolarian from the equatorial Atlantic Ocean. *Deep-Sea Res.* **27**, 715–32. (Woods Hole Oceanographic Institution cont. no. 4414.)

Thorington, G., & Margulis, L. (1981). *Hydra viridis*: Transfer of metabolites between *Hydra* and symbiotic algae. *Biol. Bull.* **160**, 175–88.

Trench, R. K. (1971). The physiology and biochemistry of zooxanthellae symbiotic with marine coelenterates. III. The effects of homogenates of host tissues on the excretion of photosynthetic products in vitro by zooxanthellae from two marine coelenterates. *Proc. Royal Soc. Lond.* Series B. 177, 251–64.

5

The *Prochloron* symbiosis

R. L. PARDY
School of Life Sciences
University of Nebraska
Lincoln, NE 68588

R. A. LEWIN
Scripps Institution of Oceanography, A-002
La Jolla, CA 92093

K. LEE
School of Life Sciences
University of Nebraska
Lincoln, NE 68588

Among those invertebrates that form symbioses with unicellular algae are several species of compound ascidians in the family Didemnidae (Eldredge, 1965). These animals are colonial, sessile filter feeders that inhabit the coastal environments of warm seas. Individual zooids making up the colony are embedded in a tough substance called *tunicine*, composed largely of cellulose. Living either within the tunicine matrix or in the common excretory tubules of the colonies are symbiotic algae. As far as is known, the algae do not inhabit the hosts' cells or internal organs. After sexual reproduction of the host, larvae are formed that in some species carry away algae in special pockets, thereby assuring the continuity of the symbiosis from one generation of hosts to another (Eldredge, 1965).

The symbionts are of interest because they possess unusual properties. Under the electron microscope, the cells look much like blue-green algae; however, they lack phycobilins and phycobilisomes, and they have chlorophyll *b* as well as *a* (Lewin & Cheng, 1975; Lewin & Withers, 1975; Newcomb & Pugh, 1975; Thorne et al., 1977). Because of these and other distinguishing features, the alga has been assigned to a separate plant division, the Prochlorophyta, in the new genus *Prochloron* (Lewin, 1976, 1977). Where the prochlorophytes fit into the evolutionary scheme of photosynthetic organisms is a matter of speculation. They may have evolved parallel with the cyanophytes. Alternatively, they may have

Part of this work was carried out aboard the R/V Alpha Helix during the Moro V Expedition and was supported by National Science Foundation Grants PCM-7904224 to R. L. Pardy, and DEB-76214 to R. A. Lewin. We thank Dr. L. Cheng for assistance in the field and help with specimen preparation. Dr. F. Lafargue is acknowledged for kindly identifying the didemnid.

arisen by mutation of a cyanophyte ancestor that developed the ability to synthesize chlorophyll *b*. It has been postulated that prochlorophytes gave rise to the chlorphytes by symbiogenesis (Lewin, 1981a, b). Existing prochlorophytes are found almost exclusively in a symbiotic existence with didemnids. To date, all attempts to bring these cells into culture have failed; contemporary prochlorophytes can thus be described as obligate symbionts.

Despite our inability to culture prochlorophytes, it has been possible to study them both in residence (*in hospite*) and isolated from some species of didemnids. Not all species of alga-bearing hosts readily yield up their symbionts. Large quantities of mucus, released by some of the animal species when the symbionts are harvested, make it almost impossible to obtain clean suspensions of the algae. In one or two species, however, fairly pure suspensions of symbionts can be obtained by gentle squeezing of the didemnid colony. Algal cells prepared in such a way are viable (capable of respiration and photosynthesis) for several hours and can be used for a variety of experiments. Analyses of intact associations and isolated algal symbionts have shown that *Prochloron* is capable of photosynthesis *in hospite* and in vitro (Tokioka, 1942; Thinh & Griffiths, 1977; Akazawa et al., 1978; Fisher & Trench, 1980). Using ^{14}C as a tracer, Pardy and Lewin (1981) found that *Prochloron* associated with *Lissoclinum patella* may release photosynthetically produced substrates to the host when *in hospite* and to the medium when in vitro. Such transfer of products from symbionts to host appears to be a universal feature of algal-invertebrate symbioses.

Among algal-invertebrate symbioses the algae usually display a high degree of physical intimacy with the host. Frequently, the symbionts reside within the hosts' cells, as in the green hydra, or in intercellular spaces enveloped by the plasma membranes of adjacent host cells, as in the green marine flatworm *Convoluta roscoffensis*. The close proximity of the symbionts to the hosts' cells probably facilitates biochemical integration of the symbionts and prevents loss of translocated products. The nature of the contact and the degree of intimacy between *Prochloron* and didemnid cells are poorly known. Although early reports (Smith, 1935; Eldredge, 1965) suggested that the algae are more or less free in the host's cloacal channels, Newcomb and Pugh (1975) published a micrograph showing *Prochloron* cells lying in cavities within the tunicine matrix of *Diplosoma virens*. The cloaca, the major excretory route, is presumably irrigated by ciliary currents, which might be expected to flush out the symbionts unless the excretory flow is low or the algae are somehow attached to the host surfaces. Here we describe some observations, made on *Prochloron* symbiotic within the didemnid *L. patella*, relevant to their possible spatial and metabolic interactions.

We collected specimens of *L. patella* from inshore water adjacent to small islets near Babelthaup Island, Republic of Palau (7°25′N, 134°30′E). Colonies, easily detached from rocks, and coral debris lying submerged at depths of 0.5–2 m were carried in seawater to the laboratory aboard the R/V Alpha Helix. Small blocks of tissue containing algal sea masses were cut out and fixed for 35 min in glutaraldehyde in phosphate-buffered seawater (pH 7.4) at 50°C. The tissues were then rinsed, postfixed with 2% osmium tetroxide in 30% sucrose for 1 h, dehydrated with ethanol and propylene oxide, and finally embedded in Epon. "Silver" sections were cut using a diamond knife in an LKB ultramicrotome.

Figure 1 (top left) shows a *Prochloron* cell, apparently free in the host's cloacal lumen, lying close to fingers of animal tissue. The algal symbiont has a central body enclosed by prominent photosynthetic membranes. This central body consists of a granular ground substance with occasional electron-dense inclusions. Not evident in this region are the fibrillar strands (thought to be DNA) reported by Schultz-Baldes and Lewin (1976) in the prochlorophytes associated with ascidians from Baja California. Intrusions of bilayered photosynthetic membranes extend into the central body from the more peripheral membranes. Thinh (1979) presented micrographs showing similar membranes in the central body of *Prochloron* cells associated with *Trididemnum cyclops*, whereas in some of the cells of *Prochloron* from *D. virens* the body appears clear and devoid of discernible organization (Thinh, 1979; cf. Whatley, 1977). Cells of *Prochloron* epizoic on ascidians collected from mangroves in Baja California exhibited little evidence of a clear central body, having instead a dense and continuous packing of subcellular structures throughout the cytoplasm (Lewin, 1975; Schultz-Baldes & Lewin, 1976). The significance of this apparent diversity is not clear. The degree to which the ultrastructure of *Prochloron* is subject to modification by the environment is unknown. Morphology of algal symbionts in green hydra is affected by the specific host strain (Pardy, 1976). Likewise, *Prochloron* morphology might be host specific.

Thylakoids are disposed in a concentric pattern around the central body of the *Prochloron* cell. Polyhedral bodies occurring among them appear similar to those shown by Schultz-Baldes and Lewin (1976). The cell wall resembles that of a typical cyanophyte and appears devoid of significant amounts of adhering material. This contrasts with the observations of Thinh (1979), whose scanning electron micrographs showed algal symbionts from *T. cyclops* covered with a sheet of fibrous material, which may serve to anchor them to the surrounding host tissue.

Although some of the symbionts in *L. patella* appear to lie free in the host's internal surface (Fig. 1 Top right, bottom left, and bottom right), some of the symbiotic cells appear completely surrounded by fingers of host tissue (Fig. 1 Bottom left and right), which may serve to hold them

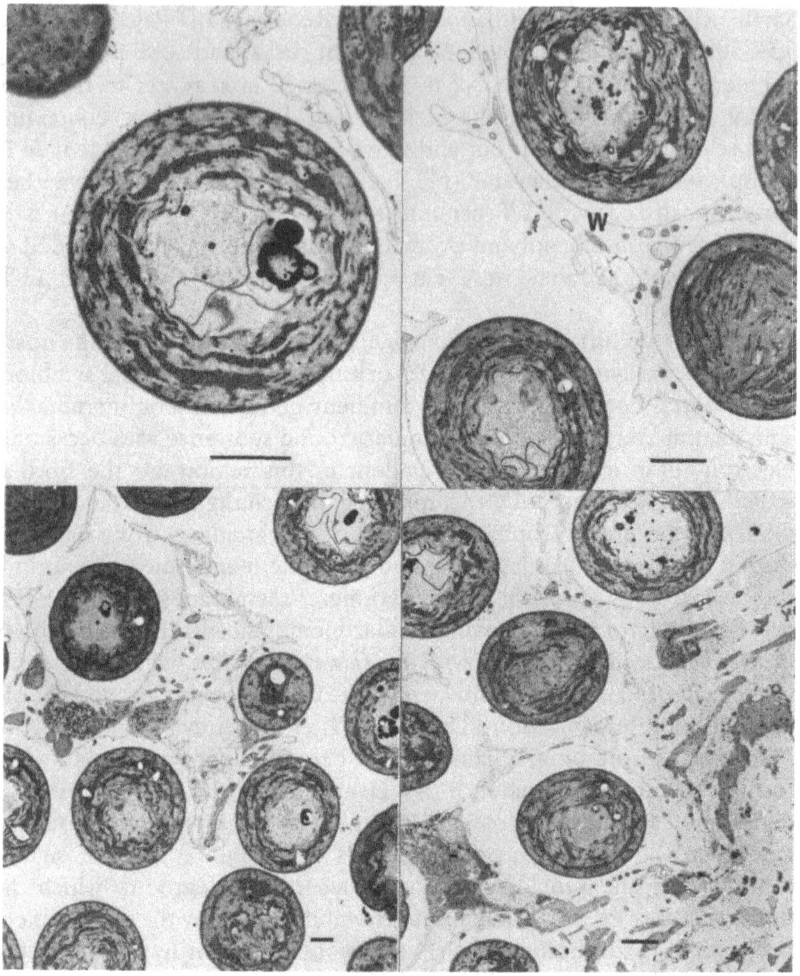

Fig. 1. Top left. An isolated *Prochloron* cell lying free in a cloacal chamber of *L. patella*, showing peripheral and intrusive thylakoids. Bar equals 1 μm. **Top right.** *Prochloron* cells lying close to an extension of a cloacal wall (W) of *L. patella*. Bar equals 1 μm. **Bottom left.** *Prochloron* cells lying both free in a cloaca and in proximity to cloacal tissue of *L. patella*. Bar equals 1 μm. **Bottom right.** *Prochloron* cells partially or completely surrounded by cloacal tissue of *L. patella*. Bar equals 1 μm.

in place. If a fresh colony of *L. patella* is removed from water, the whole colony contracts slightly and water is expelled from the common cloaca. This water is not green though it may contain symbionts. If the colony is squeezed tightly, plumes of green fluid are ejected from the cloaca. Apparently, extreme mechanical force is necessary to dislodge symbionts from the internal surfaces.

In addition to anchoring the symbiotic cells, the network of host tissue extensions that entrap the symbionts might serve as sites for metabolite exchange. Colonies of *L. patella* may acquire photosynthetic products from their *Prochloron* symbionts (Pardy & Lewin, 1981). Passage of these products from symbiont to host, as well as a possible reciprocal transfer of nutrients, may be facilitated by these sites of host-symbiont contact. The fate of free *Prochloron* cells (Fig. 1 Bottom right) is unknown. A constant, slow removal of symbionts might prevent the host from becoming overgrown with *Prochloron*. Surplus cells may be flushed out and ultimately inoculate other hosts.

The experimental harvesting of *Prochloron* symbionts from *L. patella* colonies is much easier than from any of the other four species of didemnids that we have examined, where there seems to be a closer structural intimacy between host and symbiont cells. Further study of the physical relationship between *Prochloron* and didemnids is sorely needed; in our experience, *L. patella* provides the most suitable system for such studies.

References

Akazawa, T., E. H. Newcomb, & C. B. Osmond (1978). Pathway and products of CO_2-fixation by green prokaryotic algae in the cloacal cavity of *Diplosoma virens*. *Mar. Biol.* 47, 325–30.

Eldredge, L. G. (1965). A taxonomic review of Indo-Pacific didemnid ascidians and descriptions of twenty-three central Pacific species. *Micronesia* **2**, 161–261.

Fisher, C. R., and R. K. Trench (1980). *In vitro* fixation by *Prochloron* sp. isolated from *Diplosoma virens*. *Biol. Bull.* **159**, 639–48.

Lewin, R. A. (1976). Prochlorophyta as a proposed new division of algae. *Nature* **261**, 697–8.

– (1977). *Prochloron*, type genus of the Prochlorophyta. *Phycologia* **16**, 217.

– (1981a). *Prochloron* and the theory of symbiogenesis. *Ann. N.Y. Acad. Sci.* **361**, 325–9.

– (1981b). Prochlorophytes. *In: The Prokaryotes*, ed. M. P. Starr, H. Stolp, H. G. Truper, A. Balows, & H. G. Schlegel, pp. 257–66. Berlin: Springer-Verlag.

Lewin, R. A., & L. Cheng (1975). Associations of microscopic algae with didemnid ascidians. *Phycologia* **14**, 149–52.

Lewin, R. A., & N. W. Withers (1975). Extraordinary pigment composition of prokaryotic alga. *Nature* **256**, 735–7.

Newcomb, E. H., & T. D. Pugh (1975). Blue-green algae associated with ascidians of the Great Barrier Reef. *Nature* **253**, 533–4.

Pardy, R. L. (1976). The morphology of green hydra as influenced by host strain and host environment. *J. Cell Sci.* **20**, 655–69.

Pardy, R. L., & R. A. Lewin (1981). Colonial ascidians with prochlorphyte symbionts: Evidence for translocation of metabolites from alga to host. *Bull. Mar. Sci.* **81**, 817–23.

Schulz-Baldes, M., & R. A. Lewin (1976). Fine structure of *Synechocystis didemni* (Cyanophyta: Chroococcales). *Phycologia* **15**, 1–6.

Smith, H. G. (1935). On the presence of algae in certain Ascidiacea. *Ann. Mag. Nat. Hist.* Series 10. **15**, 615–26.

Thinh, L. V. (1979). *Prochloron* (Prochlorophyta) associated with the ascidian *Trididemnum cyclops* Michaelsen. *Phycologia* **18**,77–82.

Thinh, L. V., & D. J. Griffiths (1977). Studies of the relationship between the ascidian *Diplosoma virens* and its associated microalgae. I. Photosynthetic characteristics of the algae. *Aust. J. Mar. Freshwater Res.* **28**, 673–81.

Thorne, S. W., E. H. Newcomb, & C. B. Osmond (1977). Identification of chlorophyll *b* in extracts of prokaryotic algae by fluorescence spectroscopy. *Nature* **74**, 575–8.

Tokioka, T. (1942). Ascidians found on the mangrove trees in Iwayama Bay, Palao. *Palao Tropical Biological Station Studies* **2**, 499–507.

Whatley, J. M. (1977). The fine structure of *Prochloron*. *New Phytol.* **79**, 309–13.

6

Retention of algal chloroplasts by molluscs

ROSALIND HINDE

School of Biological Sciences
The University of Sydney
New South Wales, 2006, Australia

In 1965, Kawaguti and Yamasu identified the "algal symbionts" of the sacoglossan mollusc *Elysia atroviridis* as chloroplasts, not whole algal cells. This was the first time that organelles from one species had been shown to be able to survive in the cells of another. The observation was soon followed by a considerable amount of work on the physiology of the associations, which were found in several species of *Elysia* and its relatives. These associations between sacoglossan molluscs and chloroplasts are usually referred to as *chloroplast symbioses*, but as pointed out by Trench (1980) and Hinde (1980) this term is not suitable to describe these interactions and should be replaced. Trench (1980) proposed the use of the term *foreign organelle retention* (Blackbourn et al. 1973) for all such relationships. Retention of chloroplasts is the only form of foreign organelle retention that has been observed among the Sacoglossa.

In order to determine where chloroplast retention fits into the continuum of interactions between algae and plants, we need to know the nature and degree of harm or benefit to each of the organisms. There have been a number of reviews of work on chloroplast retention (more recent ones include those of Muscatine & Greene 1973, Greene 1974, Smith 1974, Trench 1975, 1980). I will not, therefore, attempt yet another full review here. Instead, I will describe chloroplast retention in sacoglossans so as to demonstrate the probable advantages and disadvantages of the interactions involved to the plants, the chloroplasts, and the animals. I will then try to define these interactions more formally.

Establishment of the association

Sacoglossan molluscs all feed by puncturing the cells of their food and sucking out the cytoplasm, leaving the empty cell wall behind. Many

Original work described in this paper was supported by the Australian Government through the Queen's Fellowships Committee and the Australian Research Grants Committee.

species feed on coenocytic algae, particularly *Codium* and *Caulerpa* (Chlorophyta, Caulerpales) (Green 1970, Clark & Busacca 1978). Chloroplasts are taken up by the cells of the digestive gland (hepatopancreas), presumably with other components of the algal cytoplasm. However, the chloroplasts are not digested, or at least digestion is delayed. In starved animals the undigested chloroplasts continue to photosynthesize for periods ranging from a few hours or a few days in some species (e.g., *Placida dendritica*: Greene & Muscatine 1972, Hinde 1980; *Oxynoe* spp.: Clark & Busacca 1978, Hinde 1980) to between 6 and 12 weeks in others (e.g., *Elysia viridis*: Hinde & Smith 1972; *Elysia australis* and *Elysia maoria*: Hinde, unpublished observations; *Tridachia crispata, Tridachiella diomedea*: Trench et al. 1969; and *Placobranchus ianthobapsus*: Trench et al. 1969, Trench 1975). During this period, the chloroplasts continue to evolve oxygen (Kawaguti & Yamasu 1965, Trench et al. 1969, Taylor 1971a, Trench 1975, Hinde 1980) and to fix carbon dioxide in the light (Greene & Muscatine 1972, Trench et al. 1973b, Hinde 1980). The rates of photosynthesis generally appear to be considerably lower in those species that retain chloroplasts for shorter periods than in those with long-term retention (e.g., see Hinde 1980).

There is now good evidence that chloroplasts are retained for much shorter periods in slugs that are able to feed, and thus acquire fresh plastids, than they are under the stress of starvation. Gallop et al. (1980) estimated that in *E. viridis* with free access to its food plant (*Codium fragile*) one-third to one-half of the chloroplasts may be turned over each week.

The chloroplast-animal association cannot synthesize chlorophyll, glycolipids, ribulose bisphosphate carboxylase, or membrane proteins, so clearly the chloroplasts must be incapable of growth or true division when they are in the animals (Trench & Ohlhorst 1976). This makes it unlikely that chloroplasts are ever passed on from one generation of sacoglossans to the next. The eggs and veligers of a number of species of chloroplast-retaining Sacoglossa have been examined by light and electron microscopy, as well as by pigment analysis; in no case have plastids or photosynthetic pigments been found (Green 1968, Trench et al. 1969, Trench 1975, Rahat 1976, Harrigan & Alkon 1978, Clark et al. 1979, Rose, Brandley, & Hinde, unpublished). In *Elysia cauze*, chloroplasts are not incorporated until about 5 days after metamorphosis of the planktonic larvae, when the digestive gland has matured (Clark et al. 1979). *E. cauze* is the only sacoglossan in which the first incorporation of chloroplasts has been observed, but since the chloroplasts in the animals are always those of the food plant of the adult and the smallest animals found in the field always contain chloroplasts, it presumably occurs soon after metamorphosis in all species. The only sacoglossan in which the uptake of chloroplasts has actually been seen is *P. dendritica*. McLean (1976) showed that the digestive cells of the hepatopancreas take up the chloroplasts by phagocytosis, and again, this is likely to be the case in all species since molluscan digestive gland cells are commonly active in phagocytosis of food particles. In *P. dendritica*, there are signs of disruption of the chloroplasts soon after phagocytosis (within 10–20 min of the start of feeding) (McLean 1976); this is presumably the reason for the low rates of photosynthesis found in experiments with *P. dendritica* and for the relatively quick loss of photosynthetic activity.

Maintenance of chloroplasts in the digestive gland cells

In many molluscs, much of the process of digestion occurs within the digestive cells of the hepatopancreas. These cells can digest chloroplasts, since degraded chloroplasts are seen from time to time in electron micrographs of sacoglossans (Hinde, unpublished). In *E. viridis* (Trench et al. 1973b) and in *E. australis* and *E. maoria* (Hinde, unpublished) that have been starved for several weeks, an unusually high proportion of the chloroplasts are seen to be in various stages of digestion. In all cases where degradation of chloroplasts has been observed, the chloroplasts are surrounded by an obvious membrane, outside the chloroplast envelope, which must have been produced by the animal cell. There is some conflict in the literature about the existence of host-produced membranes around the apparently normal chloroplasts in slugs that have not been starved. Trench et al. (1973b) observed that about half the chloroplasts in *E. viridis* lay free in the cytoplasm, while the others were surrounded by a host-produced membrane. Hawes (1979), also working with *E. viridis*, found that there was a host-produced membrane around each chloroplast; he regarded this membrane as homologous with the membrane of the original phagocytic vacuole. In thin sections of *Elysia chlorotica* (Graves et al. 1979) and of *E. australis* (Hinde, unpublished), and in sections and freeze-fractured material from *E. maoria* and from *Elysia* cf. *furvacauda* (Brandley 1981, 1982), the majority of the chloroplasts appear to lie free in the cytoplasm. The disappearance of the phagosome membrane may be important in protecting the chloroplasts from digestion, since lysosomal enzymes are normally released only after fusion of the membranes of a lysosome and a phagosome (or an autophagic vacuole). After they have been engulfed by macrophages, some pathogens of mammals avoid digestion by escaping from the phagosome; the mechanism of escape is not yet understood (Hart 1979). Possibly chloroplasts that are still photosynthesizing provide some sort of signal to the digestive gland cell, which inhibits their destruction (Brandley 1982).

As pointed out, the members of the order Sacoglossa vary greatly in their ability to retain chloroplasts, some being incapable of using ingested chloroplasts at all, others maintaining chloroplast function briefly or for very long periods. Generally, the species of the family Elysiidae (the most advanced family of the order) maintain their chloroplasts for the longest periods. All this indicates that various adaptations are required for efficient exploitation of retained chloroplasts, and that these have evolved during the evolution of the Sacoglossa.

Most of the published investigations of chloroplast retention by sacoglossans have dealt with species that obtain their chloroplasts from siphonaceous green algae from the orders Caulerpales and Codiales. Because of this, it has been suggested that it is the unusual properties of these chloroplasts that prevent digestion of the plastids after phagocytosis. The chloroplasts of many siphonaceous (or coenocytic) green algae are known to be unusually resistant to osmotic shock and to the stresses imposed by isolation and by storage outside the plant. For example, isolated chloroplasts from *C. fragile* fix carbon over a wide range of osmotic pressures. After 7 days' storage at 9°C in the dark they have lost only 22% of their chlorophyll, and at this time they still fix $^{14}CO_2$ at 20% of the rate observed immediately after isolation (Gallop et al. 1980). It has been suggested by several authors that only such "robust" chloroplasts could maintain their photosynthetic activity after incorporation into the digestive cells of an animal (see Trench 1975). There is now good evidence that chloroplasts from nonsiphonaceous Chlorophyta (Hawes 1979, Hinde 1980, Brandley 1982), from the coenocytic alga *Vaucheria* (Xanthophyta) (Hinde & Smith 1974, Graves et al. 1979, West 1980), and from red algae (Rhodophyta) (Taylor 1971b, Kremer & Schmitz 1976, Brandley 1982) may be retained and function in the digestive cells of sacoglossans of various species. It is unlikely that chloroplasts from all these sources share the "robust nature" of the plastids of *Codium* and *Caulerpa*. Thus, it appears most likely that it is adaptations of the sacoglossans, rather than properties of the chloroplasts, that make it possible for the molluscs to take up undamaged chloroplasts and maintain them in the digestive gland cells in a state in which they can still photosynthesize (see also Hinde 1980). The fact that there are many species of Sacoglossa that feed on siphonaceous green algae but do not retain chloroplasts (Green 1970, Hinde & Smith 1974) also suggests that it takes more than the ingestion of robust chloroplasts from one of the Caulerpales or Codiales by a sacoglossan to establish a chloroplast-mollusc association. The range of chloroplast types that can photosynthesize in the cytoplasm of an animal cell can be extended, since chloroplasts from a member of either the Bacillariophyceae or the Chrysophyceae have been found in certain Foraminifera (Lopez 1979), and chloroplasts resembling those of the Dinophyceae, Haptophyceae, and Chrysophyceae have been found in two species of the Ciliophora (Blackbourn et al. 1973).

Apart from the adaptations needed to avoid digestion of the chloroplasts and those that presumably ensure maintenance of their photosynthetic ability (e.g., by supply of metabolites), sacoglossans show a second type of adaptation to their associations with retained chloroplasts: They produce a "host factor" that causes products of photosynthesis to leak from the chloroplasts (Gallop 1974). Chloroplasts isolated from *C. fragile* normally lose only a small proportion (1%–20%, depending on experimental conditions) of freshly fixed carbon into the suspension medium in the first hour after a pulse of fixation of $^{14}CO_2$; most of the ^{14}C released is in glycollate (Trench et al. 1973a, Gallop 1974). Under the same conditions, homogenates of *E. viridis* strongly stimulate movement of fixed carbon to the medium, with as much as 73% being released in the presence of the homogenate (Gallop 1974). The effect is specific in that only certain compounds move out of the chloroplasts. Glycollate, glucose and alanine were the only labeled compounds recovered from the media in these experiments, although the chloroplasts contained many other labeled compounds (Gallop 1974). Gallop (1974) also demonstrated that homogenates of parts of the animals that do not contain chloroplasts (i.e., tissues other than the digestive gland) did not stimulate transport of photosynthetic products out of the chloroplasts, and neither did homogenates of *C. fragile*. The evolution of the ability to retain chloroplasts has apparently involved evolution of a mechanism to ensure translocation of a large proportion of the carbon fixed during photosynthesis to the cytoplasm of the animals' cells. This mechanism appears to be distinct from that involved in movement of photosynthate from the plastids to the cytoplasm in the plant.

Benefits and harm to the organisms and plastids involved in chloroplast retention

The animals

Much of the carbon fixed photosynthetically by the chloroplasts moves into animal cells where it is metabolized by a number of pathways. For example, in *T. crispata*, 30% of all the carbon fixed during a 1-h period of exposure to $^{14}CO_2$ in the light was recovered from mucus after a chase period of 5 h (Trench et al. 1972). In *E. viridis*, 36% of the ^{14}C fixed during a 15-min "pulse" exposure to $^{14}CO_2$ was converted to galactose by the animal tissue (neither chloroplasts isolated from *C. fragile* nor the intact algae synthesize galactose from freshly fixed carbon) (Trench et al. 1973b). Carbon-14 that has been fixed photosynthetically in *E. viridis* can be recovered from proteins and lipids, as well as from carbohydrates, over a 4-day period after its incorporation (Hinde 1978). Since the rate of photosynthesis of normal *E. viridis* is comparable with that of *C. fragile* (Trench et al. 1973b) and since at least 36% of photosynthetically fixed carbon is eventually metabolized by the animals' cells, it seems likely that in *E. viridis*, at least, photosynthesis is an important source of organic nutrients for the animals. *E. viridis* deprived of both light and food lose weight approximately twice as fast as those deprived only of food (Hinde & Smith 1975). In other words, *E. viridis* depends on photosynthesis for an important part of its nutrition. However, in most cases chloroplast-retaining sacoglossans are not fully autotrophic. *Elysia* species, and many other Sacoglossa, are usually found living among their food plants, and they appear to feed frequently. Because the chloroplast population is turned over quite rapidly (Clark & Busacca 1978, Gallop et al. 1980), the animals probably have to feed at intervals simply to maintain their supply of functional chloroplasts. However, experiments in which the effects of depriving the animals of either food or photosynthesis were measured suggest that feeding is also important as a source of nutrients in *E. viridis* (Hinde & Smith 1975) and *E. australis* (Hinde, unpublished). The frequency of feeding in many sacoglossan species seems to be greater than would be needed if its only function were the replacement of chloroplasts.

Since the animals must feed to replenish their chloroplasts, they may obtain part of their energy needs from their food. There must be metabolic costs involved in maintaining the chloroplasts, so the animals almost certainly gain advantages, other than a simple flow of metabolites for energy production, from having the chloroplasts. At present we do not know what these advantages are. On the other hand, because many chloroplast-bearing species fix carbon at rates comparable with those of their food plants, it is necessary to ask why they need to feed at all, once they have chloroplasts. Before we can fully define the relationship between these sacoglossans, their food algae, and the chloroplasts, we will need to know much more about why the animals require both food and photosynthesis. There are a number of possibilities, which are not mutually exclusive:

1. Possession of plastids provides excellent camouflage, at least while the animal remains on its food plant. Little is known of the nature of possible predators of sacoglossans or of their significance in the biology of the animals, and so the real importance of camouflage to these animals cannot be assessed.

2. Mutualistic symbioses between microorganisms and animals often allow the animal species to exploit food sources that lack particular essential nutrients (e.g., in mammals various vitamins are synthesized by bacteria in the gut), or even to live on diets of extraordinarily limited composition (as with termites, which can survive and reproduce on diets of pure cellulose). The diets of herbivorous sacoglossans may be incomplete or unbalanced in comparison with those of other herbivorous molluscs, since sacoglossans eat the cytoplasm but not the cell walls of their

food plants, whereas most herbivorous molluscs do eat the cell wall and can digest at least some of its constituent carbohydrates (Owen 1966). This suggests that the carbon: nitrogen ratio of the diet may be too low, particularly as the sacoglossans produce large amounts of mucus and polysaccharide jelly for their egg masses. Use of proteins from food as a source of energy or for conversion to carbohydrates is inefficient, and the production of carbohydrates during photosynthesis might allow these proteins to be used more efficiently for growth. Isolated chloroplasts can assimilate inorganic nitrogen (nitrate, nitrite and ammonia), forming organic nitrogen compounds (Morris 1974), but there has been no investigation of this aspect of the metabolism of chloroplasts in any of the sacoglossans that retain them. Carbon-labeled amino acids are produced during photosynthesis in sacoglossans that retain chloroplasts (Trench et al. 1973b, Hinde 1978). This may indicate assimilation of nitrogen by the chloroplasts, but since conditions in the animal cells (e.g., poor availability of inorganic nitrogen sources or high levels of organic nitrogen compounds) are probably quite unlike those in the plant, the rates of assimilation probably differ from those in the plant, and may be much lower.

3. If the chloroplasts themselves can supply a balanced "diet" of organic carbon and organic nitrogen to the animals, the animals may still depend on eating whole algal tissue, because they require one or more organic compounds neither they nor the chloroplasts can synthesize (e.g., vitamins, essential fatty acids, or essential amino acids).

The question of why both feeding and photosynthesis are needed by Sacoglossa that retain chloroplasts can only be answered by a thorough investigation of the nutritional physiology of sacoglossans, with and without chloroplasts, and of the ecology of these animals.

The chloroplasts

Retention by sacoglossans cannot be said to be of any real benefit to the chloroplasts since they are not autonomous, self-reproducing organisms. They do reproduce by dividing in plant cells, but much of the information needed for their repair, growth and division is located in the plant cell nucleus. As mentioned earlier, Trench and Ohlhorst (1976) have shown that in *T. crispata* and *E. viridis* the chloroplasts are incapable of synthesizing a number of their own important constituents. These results confirm their lack of genetic autonomy.

The plants

The relationship of the sacoglossan to the alga from which it derives its chloroplasts is that of a browser to its food. The fact that the chloroplasts do not "die" immediately after the animal has damaged the plant by feeding on it is not an extra disadvantage, but it can be of little or no benefit to the plant either. As soon as they are ingested by the animal the chloroplasts are on the way to being digested, even if digestion is delayed for several months. The productivity of the chloroplasts while they are in the animals may decrease the amount of plant material the animals eat. To this extent the ability of the chloroplasts to survive in animal cells may slightly enhance the fitness of the plants from which they come. Further work is needed to assess the impact of grazing by sacoglossans on the populations of their food plants, and thus the degree to which natural selection may have led to the "robust" properties of chloroplasts in some algae. However, most sacoglossans are small animals and they are usually rare; those I have worked with are sparsely distributed on their food plants and cause little visible damage.

Concluding remarks: the nature of the association between sacoglossans, their food plants, and their retained chloroplasts

The relationship between the sacoglossans and their retained chloroplasts has, until recently, been referred to as *chloroplast symbiosis*. Since the term *symbiosis* is so often used in the restricted sense of "mutualism," use of the word has tended to suggest that chloroplast retention is a mutualistic relationship. The research outlined in this chapter has shown without doubt that the association is beneficial to the animals harboring the chloroplasts. Since symbioses are relationships between organisms of different species, and chloroplasts are not genetically autonomous organisms, chloroplast retention does not fall within the range of symbioses at all, if the chloroplast and the animal are considered the partners in the relationship. The definition of mutualistic symbiosis implies that the benefits to the partners will increase their Darwinian fitness, and clearly this is not possible if one of the members of the association is not an entity capable of independent evolution. If the plant from which the chloroplasts are obtained is considered to be the other member of the association, it is immediately clear that the relationship is simply a feeding one, not a mutualistic symbiosis. The animals feed on the algae, causing damage that must be repaired and that may, therefore, reduce the reproductive potential of the plants. There may have been some selective pressure on the food plants tending to produce chloroplasts that could survive longer in the Sacoglossa, thus reducing grazing on the algae.

Summary

Many species of sacoglossan mollusc retain intact, photosynthesizing chloroplasts in the cells of their digestive glands. The establishment and maintenance of these associations are described, and the harm and benefit to the partners are discussed. The algae eaten by these sacoglossans are food for the animals, and thus they are harmed by the association. The animals are browsers on the algae and gain both from feeding on the plants and from the presence of the retained chloroplasts. The chloroplasts are not genetically autonomous and therefore cannot be regarded as being benefited or harmed by their incorporation into animal cells. It is possible that selection pressure due to grazing by sacoglossans has produced chloroplasts capable of unusually extended survival in animal cells, thus decreasing grazing on the plants. It is clear that the animals have very specialized strategies that allow them to exploit the chloroplasts.

References

Blackbourn, D. J., Taylor, F. J. R., & Blackbourn, J. (1973). Foreign organelle retention by ciliates. *J. Protozoo.* **20**, 286–8.

Brandley, B. K. (1981). Ultrastructure of the envelope of *Codium australicum* (Silva) chloroplasts in the alga and after acquisition by *Elysia maoria* (Powell). *New Phytol.* **89**, 679–86.

– (1982). "Investigations of Mechanisms of Chloroplast Retention in Elysiid Sacoglossans." Ph.D. thesis, University of Sydney.

Clark, K. B., & Busacca, M. (1978). Feeding specificity and chloroplast retention in four tropical Ascoglossa, with a discussion of the extent of chloroplast symbiosis and the evolution of the order. *J. Molluscan Studies* **44**, 272–82.

Clark, K. B., Busacca, M., & Stirts, H. (1979). Nutritional aspects of development of the ascoglossan, *Elysia cauze. In: Reproductive Ecology of Marine Invertebrates*, ed. S. Stancyk, pp. 11–24. Belle W. Baruch Library in Marine Science No. 9. Columbia: University of South Carolina Press.

Gallop, A. (1974). Evidence for the presence of a "factor" in *Elysia viridis* which stimulates photosynthate release from its symbiotic chloroplasts. *New Phytol.* **73**, 1111–17.

Gallop, A., Bartrop, J., & Smith, D. C. (1980). The biology of chloroplast acquisition by *Elysia viridis*. *Proc. Royal Soc. Lond.* Series B. **207**, 335–49.

Graves, D. A., Gibson, M. A., & Bleakney, J. S. (1979). The digestive diverticula of *Alderia modesta* and *Elysia chlorotica* (Opisthobranchia: Sacoglossa). *Veliger* **21**, 415–22.

Greene, R. W. (1968). The egg masses and veligers of southern California sacoglossan opisthobranchs. *Veliger* **11**, 100–4.

– (1970). Symbiosis in sacoglossan opisthobranchs: symbiosis with algal chloroplasts. *Malacologia* **10**, 357–68.

– (1974). Sacoglossans and their chloroplast endosymbionts. *In: Symbiosis in the Sea*, ed. W. B. Vernberg, pp. 21–7. Belle W. Baruch Library in Marine Science No. 2. Columbia: University of South Carolina Press.

Greene, R. W., & Muscatine, L. (1972). Symbiosis in sacoglossan opisthobranchs: Photosynthetic products of animal-chloroplast associations. *Mar. Biol.* **14**, 253–9.

Harrigan, J. F., & Alkon, D. L. (1978). Laboratory cultivation of *Haminoea solitaria* (Say, 1822) and *Elysia chlorotica* (Gould, 1870). *Veliger* **21**, 299–305.

Hart, P. d'A. (1979). Phagosome-lysosome fusion in macrophages: A hinge in the intracellular fate of ingested microorganisms. *In: Lysosomes in Applied Biology and Therapeutics*, vol. 6, ed. J. T. Dingle, P. J. Jacques, & I. H. Shaw, pp. 409–23. Amsterdam: North-Holland.

Hawes, C. R. (1979). Ultrastructural aspects of the symbiosis between algal chloroplasts and *Elysia viridis*. *New Phytol.* **83**, 445–50.

Hinde, R. (1978). The metabolism of photosynthetically fixed carbon by isolated chloroplasts from *Codium fragile* (Chlorophyta: Siphonales) and by *Elysia viridis* (Mollusca: Sacoglossa). *Biol. J. Linn. Soc.* **10**, 329–42.

– (1980). Chloroplast "symbiosis" in sacoglossan molluscs. *In: Endocytobiology: Endosymbiosis and Cell Biology, a Synthesis of Recent Research*, ed. W. Schwemmler & H. E. A. Schenk, pp. 729–36. Berlin: Walter de Gruyter.

Hinde, R., & Smith, D. C. (1972). Persistence of functional chloroplasts in *Elysia viridis* (Opisthobranchia, Sacoglossa). *Nat. New Bio.* **239**, 30–1.

– (1974). "Chloroplast symbiosis" and the extent to which it occurs in Sacoglossa (Gastropoda: Mollusca). *Biol. J. Linn. Soc.* **6**, 349–56.

– (1975). The role of photosynthesis in the nutrition of the mollusc *Elysia viridis*. *Biol. J. Linn. Soc.* **7**, 161–71.

Kawaguti, S., & Yamasu, T. (1965). Electron microscopy on the symbiosis between an elysioid gastropod and chloroplasts from a green alga. *Biol. J. of Okayama Univ.* **11**, 57–65.

Kremer, B. P., & Schmitz, K. (1976). Aspects of $^{14}CO_2$-fixation by endosymbiotic rhodoplasts in the marine opisthobranchiate *Hermaea bifida*. *Mar. Biol.* **34**, 313–16.

Lopez, E. (1979). Algal chloroplasts in the protoplasm of three species of benthic Foraminifera: taxonomic affinity, viability and persistence. *Mar. Biol.* **53**, 201–11.

McLean, N. (1976). Phagocytosis of chloroplasts in *Placida dendritica* (Gastropoda: Sacoglossa). *J. Exper. Zoo.* **197**, 321–30.

Morris, I. (1974). Nitrogen assimilation and protein synthesis. *In: Algal Physiology and Biochemistry*, ed. W. D. P. Stewart, pp. 583–609. Oxford: Blackwell Scientific Publications.

Muscatine, L., & Greene, R. W. (1973). Chloroplasts and algae as symbionts in molluscs. *Intern. Rev. of Cytology* **36**, 137–69.

Owen, G. (1966). Digestion. *In: Physiology of Mollusca*, vol. 2, ed. K. M. Wilbur & C. M. Yonge, pp. 53–96. New York: Academic Press.

Rahat, M. (1976). Direct development and symbiotic chloroplasts in *Elysia timida* (Mollusca: Opisthobranchia). *Israel J. of Zoo.* **25**, 186–93.

Smith, D. C. (1974). Transport from symbiotic algae and symbiotic chloroplasts to host cells. *Symp. Soc. Exper. Biol.* **28**, 485–520.

Taylor, D. L. (1971a). Photosynthesis of symbiotic chloroplasts in *Tridachia crispata* (Bërgh). *Comp. Biochem. Phys.* **38A**, 233–6.

– (1971b). Symbiosis between the chloroplasts of *Griffithsia flosculosa* (Rhodophyta) and *Hermaea bifida* (Gastropoda: Opisthobranchia). *Pubbli. Staz. Zoo. Nap.* **39**, 116–20.

Trench, R. K. (1975). Of "leaves that crawl": functional chloroplasts in animal cells. *Symp. Soc. Exper. Biol.* **29**, 229–65.

– (1980). Uptake, retention and function of chloroplasts in animal cells. *In: Endocytobiology: Endosymbiosis and Cell Biology, a Synthesis of Recent Research*, ed. W. Schwemmler & H. E. A. Schenk, pp. 703–27. Berlin: Walter de Gruyter.

Trench, R. K., & Ohlhorst, S. (1976). The stability of chloroplasts from siphonaceous algae in symbiosis with sacoglossan molluscs. *New Phytol.* **76**, 99–109.

Trench, R. K., Greene, R. W., & Bystrom, B. G. (1969). Chloroplasts as functional organelles in animal tissues. *J. Cell Bio.* **42**, 404–17.
Trench, R. K., Trench, M. E., & Muscatine, L. (1972). Symbiotic chloroplasts; their photosynthetic products and contribution to mucus synthesis in two marine slugs. *Biol. Bull.* **142**, 335–49.
Trench, R. K., Boyle, J. E., & Smith, D. C. (1973a). The association between chloroplasts of *Codium fragile* and the mollusc *Elysia viridis*. I. Characteristics of isolated *Codium* chloroplasts. *Proc. Royal Soc. Lond.* Series B. **184**, 51–61.
– (1973b). The association between chloroplasts of *Codium fragile* and the mollusc *Elysia viridis*. II. Chloroplast ultrastructure and photosynthetic carbon fixation in *E. viridis*. *Proc. Royal Soc. Lond.* Series B. **184**, 63–81.
West, H. H. (1980). Photosynthetic oxygen evolution by symbiotic chloroplasts in molluscan host. *Amer. Zoo.* **20**, 798.

7

The *Azolla-Anabaena azollae* symbiosis

GERALD A. PETERS AND HARRY E. CALVERT
Charles F. Kettering Research Laboratory
Yellow Springs, OH 45387

The photosynthetic prokaryotes commonly referred to as blue-green algae are now more commonly termed *cyanobacteria* (Haselkorn, 1978). Morphologically this group of organisms includes unicellular, pleuorcapsalean, and filamentous forms, the latter including simple, branched, and multiseriate filaments. Some of the filamentous forms are distinguished by their ability to differentiate two specialized cell types: heterocysts and akinetes. Numerous forms of interaction have been described between cyanobacteria and other prokaryotic as well as eukaryotic organisms (Whitton, 1973). However, the occurrence of cyanobacteria in well-defined symbiotic associations is relatively rare and restricted to only a few genera.

Lichens, comprised of an ascomycete fungus (mycobiont) and a green alga, a cyanobacterium, or both, are perhaps the most familiar and diverse group of symbiotic associations. Although there are some 500 genera and 17,000–18,000 species of lichens, only 8% of them contain a cyanobacterium as a phycobiont (Fogg et al., 1973; Millbank, 1974, 1977). Furthermore, though eight genera of cyanobacteria are represented, *Nostoc* species are by far the most common. The number of genera is even more restricted in the green plant-cyanobacterium symbioses. Such associations occur with representatives from a broad segment of the plant kingdom, including several bryophytes (*Anthoceros, Blasia, Cavicularia*), a pteridophyte (*Azolla*), gymnosperms (nine genera of cycads, including *Cycas, Zamia, Macrozamia*, and *Encephalartos*), and an angiosperm (*Gunnera* in the Halagoraceae). However, it is important to note that the actual number of plant genera is very small. Moreover, the cyanobacterium in the symbiosis is always a member of the Nostocaceae, a family charac-

The authors are indebted to Mr. R. E. Toia, Jr. and Mr. M. K. Pence for both technical assistance and aid in preparation of this manuscript. We also thank Mr. S. R. Dunbar and Mrs. M. Z. Tootle for photographic assistance, Mrs. D. Patten for clerical assistance, and Dr. W. D. Bauer for critically reading the manuscript. Studies described here that originated in the authors' laboratories were supported in part by NSF grants DEB 74-11679 AO1 and PFR 77-27269, and USDA-SEA grant 5901-0410-9-0330-0.
Contribution No. 741 from the C. F. Kettering Research Laboratory.

terized by the ability of its members to differentiate heterocysts and fix atmospheric N_2. Thus, in these plant-cyanobacterium symbioses, as in the legume-*Rhizobium* symbioses and the actinorhizal symbioses, the prokaryotic organism provides the eukaryotic plant with a utilizable source of nitrogen.

The occurrence of the plant-cyanobacteria symbioses has been documented for 50–100 years, but until the early 1970s they were for the most part simply considered botanical curiosities. During the past decade each of these associations has been "rediscovered" and studied in varying degrees with modern research technology. This renewed interest has been based primarily on a single factor: their capacity to utilize atmospheric N_2 for growth. Modern agriculture is extremely dependent upon commercial fertilizer nitrogen to maintain high crop productivity. The production of commercial fertilizer nitrogen is, however, based on the energy-intensive Haber-Bosch process in which N_2 is converted to ammonia using a metal catalyst, temperatures of about 400° C, pressure of 300 atm, and a source of H_2, usually natural gas. Therefore, the diminishing availability and/or questionable stability of energy sources and their increasing costs necessarily have an impact upon the production of commercial fertilizer nitrogen. This in turn affects world food production. Among the crop plants, only the nodulated legumes do not require fertilizer nitrogen for high productivity. This is due to their symbiosis with *Rhizobium*, one of a small number of prokaryotic organisms, including some cyanobacteria that contain the O_2-labile enzyme, nitrogenase. Nitrogenase catalyzes the reduction of N_2 to ammonia.

During the past decade there has been a worldwide stimulation of research on biological N_2 fixation and interest in its potential for alleviating the dependence of agriculture on fossil fuel energy-dependent nitrogen sources. It is in this framework that we begin our consideration of the *Azolla-Anabaena azollae* associations.

Historical perspective

Among the known plant-cyanobacterial symbioses, only the *Azolla-Anabaena* associations have significant potential as an alternative nitrogen source in agriculture (Moore, 1969; Lumpkin & Plucknett, 1980). These associations have a long history of use as a green compost for rice in the Far East. According to Liu (1979), a book on agricultural techniques written by Jia Ssu Hsieh in 540 A.D., entitled *The Art of Feeding the People* (*Chih Min Tao Shu*), describes the cultivation and use of *Azolla* in rice fields. At the end of the Ming dynasty (beginning of the seventeenth century) there were many "local" records of its use as a green compost. However, the use of *Azolla* in the Peoples' Republic of China did not begin to expand markedly until 1962 (Liu, 1979). Its use currently has been extended to approximately 1.34 million hectares (ha). In Vietnam, the use of *Azolla* can likewise be traced back many centuries and there are several legends about its domestication (Moore, 1969; Dao & Tran, 1979; Lumpkin & Plucknett, 1980), one of which is that the cultivation of *Azolla* was promoted by a Buddist monk, Khong Minh Khong, in the eleventh century (Dao & Tran, 1969). In 1955, the use of *Azolla* was still rather limited, restricted to about 40,000 ha in the Red River Delta. Subsequently, techniques for its multiplication were studied and popularized, as was its dual cropping with winter rice, and its use was extended to 320,000 ha by 1965 (Dao & Tran, 1979). Although the use of *Azolla* in Vietnam has declined slightly at present (Dao The Tuan, personal communication, IRRI workshop, 1980), about 40% of the rice field area in northern Vietnam was used for *Azolla* culture after the summer rice harvest in 1976 (Watanabe, 1978).

In most developing countries rice is a major source of food calories, especially among the poor, and thus it is a pivotal crop in efforts to stave off famine. Since nitrogen is most often the nutrient limiting crop yield, especially with new high-yield varieties, the use of *Azolla* as an alternative nitrogen source is now receiving considerable interest in other countries. Detailed field studies were initiated during 1975–6 at the International Rice Research Institute (IRRI) in the Philippines (Watanabe et al., 1977), at the Central Rice Research Institute (CRRI), Cuttack, India (Singh, 1977), and at the University of California, Davis (Talley et al., 1977) to determine further the problems and the applicability of *Azolla* in rice production. These continuing studies (Singh, 1979, 1980; Subudhi & Watanabe, 1979; Talley & Rains, 1980a, b; Watanabe et al., 1980) have clearly demonstrated the potential of *Azolla* as an alternative or supplemental nitrogen source for rice in tropical and temperate regions. *Azolla* is most effective when grown as a green compost during the fallow season of rice and incorporated into the soil, but nitrogen is also provided when *Azolla* is grown in dual culture as a cover crop with rice (Talley et al., 1977). Although *Azolla* use is labor intensive, as much as 100 kg N/ha can be provided to a single rice crop at IRRI in the Philippines using the Chinese dual row method (Fig. 1) and, after briefly draining the paddy, incorporating the *Azolla* into the soil at intervals throughout the growing season. In California, 60 kg N/ha appears to be a reasonable and perhaps slightly conservative estimate of the nitrogen input that can be expected from an *Azolla* crop grown during the fallow season (Talley & Rains, 1980b; Rains, personal communication).

Azolla also has a long history of use as a fodder. Moore (1969) cites reports of its use with or without other aquatic plants as a feedstock for pigs and ducks. Chinese farmers have long used *Azolla* as a "fresh, cooked or fermented" fodder for pigs because a hectare of *Azolla* can provide enough fodder for 150–300 pigs (Liu, 1979). In Vietnam, where co-

Fig. 1. *Azolla pinnata* growing in dual culture with rice at the International Rice Research Institute. The rice has been planted in the double narrow rows used for the continuous cultivation and incorporation of *Azolla* during growth of the rice crop in the Peoples' Republic of China.

operatives grow *Azolla* for animal food, it is fed to pigs and, to a lesser extent, to cattle and poultry (Dao & Tran, 1979). *Azolla* can also be used as a fish food (Sculthorpe, 1967; Lumpkin & Plucknett, 1980).

Azolla species are subject to a number of pests, some of which can be disastrous to successful propagation if they are not effectively controlled. Liu (1979) describes the characteristics and life habits of the major pests of *Azolla* in the Peoples' Republic of China as well as control methods, which include an integration of chemical and biological approaches. Singh (1977, 1979) and Watanabe (1978) both note attacks by insects, specifically larvae of the Lepidoptera (*Nymphula*) and Diptera (*Chironomus*) and their control with carbofuran. Other pests, including water mites and aphids, are noted in Lumpkin and Plucknett (1980). In general, the insect pests of *Azolla* are distinct from those of rice but can be controlled by the same insecticides.

Taxonomy and sexual reproduction

Azolla is a genus of aquatic heterosporous ferns established by J. B. Lamarck in 1783 (Svenson, 1944). The genus is usually included with

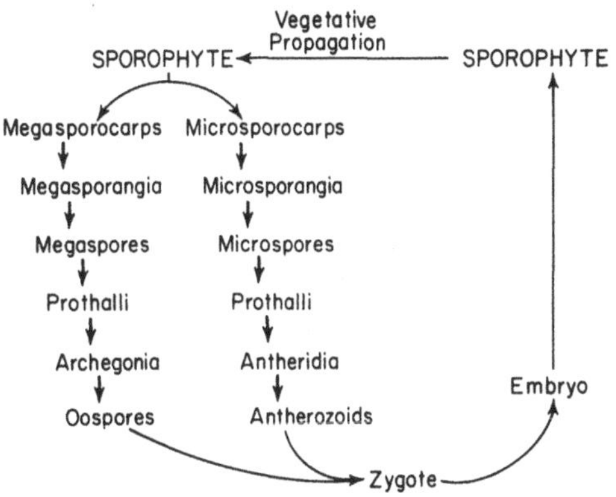

Fig. 2. Life cycle of *Azolla*.

Salvinia in the Salviniaceae (Smith, 1955). However, in recent years some authors have placed *Azolla* in a separate family, the Azollaceae (see Sculthorpe, 1967; Moore, 1969; Konar & Kapoor, 1974; Lumpkin & Plucknett, 1980). Species demarcation is based primarily upon reproductive structures. The sexual cycle and/or sporocarp development has been described in considerable detail for *A. filiculoides* and *A. pinnata* and to a lesser extent for other *Azolla* species (see Lumpkin & Plucknett, 1980, for specific references; Lucas & Duckett, 1980, for cytological details). The sexual cycle enables the plants to overwinter and withstand desiccation.

A generalized life cycle for *Azolla* is shown in Fig. 2. Sporocarps are borne in pairs (tetrads in *A. nilotica*) on short stalks that arise from the first ventral leaf lobe initial of a lateral branch. The sporocarp pair may comprise two microsporocarps (male), two megasporocarps (female), or one of each. Microsporocarps are appreciably larger than the megasporocarps. During sporocarp development the megaspore mother cell derived from the ventral lobe initial divides, ultimately producing 32 megaspore nuclei. At this stage either all but one of the nuclei abort, with the survivor giving rise to a megaspore, or they all abort and microsporangial initials arise from basal outgrowths on the stalk of the megasporangium. A megasporocarp contains a single megasporangium with one megaspore and the surrounding megaspore apparatus, which includes floats and capture mechanism. Mature microsporocarps may contain anywhere from 8 (Svenson, 1944) to 130 (Duncan, 1940) stalked microsporangia, each of which may develop 32 or 64 microspores aggregated into three to ten massulae. (Each massula consists of a mass of microspores embedded in a mucilaginous matrix from the sporangial wall.)

Mature mega- and microsporocarps dehisce from the sporophyte. The

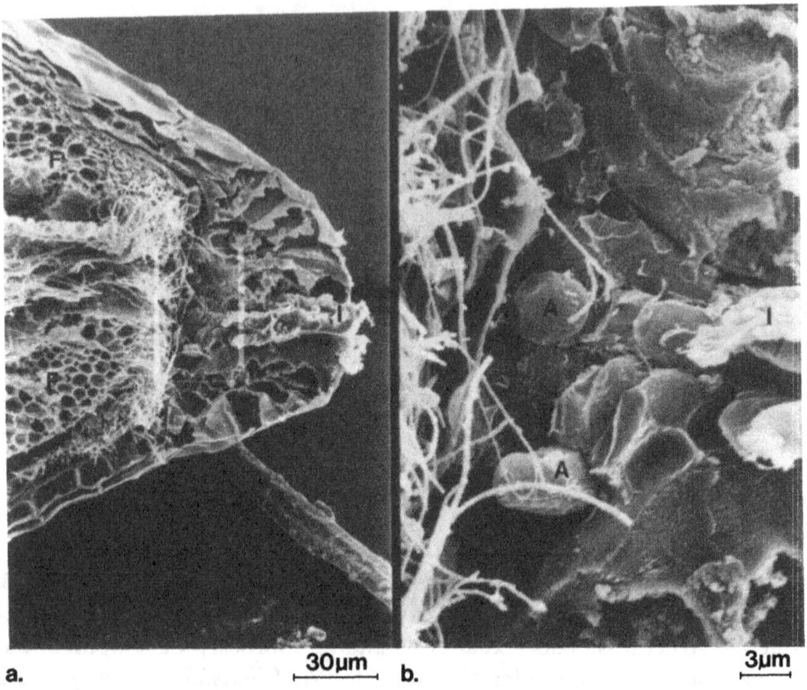

Fig. 3. The endophytic *Anabaena* in a megasporocarp of *Azolla mexicana* Presl. (a) Tip of a dissected megasporocarp showing the indusium (I) and portions of two floats (F). (b) Enlargement of the region between the indusium and floats which contains *Anabaena* cells (A).

microsporocarps disintegrate, releasing massulae with glochidia (except for *A. nilotica*). The glochidia are barbed bristles that serve to anchor the massulae to the megaspores. The megaspores germinate into female gametophytes, each of which produces one or more archegonia. The microspores germinate into male prothalli, which differentiate antheridia and release antherozoids. Fertilization takes place either underwater (Bierhorst, 1971) or on wet surfaces of gametophytes in aggregations of massulae and megasporocarps on the leeward side of ponds or paddies (S. Talley & P. K. Singh, personal communications). Subsequent development of the embryo into the mature sporophyte has been described in detail for *A. pinnata* (Konar & Kapoor, 1974).

Akinetes of *Anabaena* are apparently present during the development of both the mega- and microsporocarps but are retained to maturity only in the former (Fig. 3). There has been some confusion on this point in the literature in that some authors have reported *Anabaena* spores in the microsporocarps whereas others have not. The discrepancy is undoubtedly due to the developmental stage of the material observed. During the development of the zygote the *Anabaena* akinetes germinate to produce undifferentiated, generative filaments. These filaments become associated with the shoot apex of the developing sporophyte, perpetuating the symbiosis through the reproductive cycle. The continuity of the *Anabaena* endophyte throughout the life cycle of the *Azolla* does not preclude the possibility of a free-living form of the endophyte. However, it does eliminate the necessity for a free-living form. In this regard these associations are distinct from all other plant-prokaryote symbioses.

Extant species of *Azolla* are divided into two sections, based on the number of megaspore floats (Florschütz, 1938; Svenson, 1944; Moore, 1969). These are (1) the *Euazolla* (three floats), which is currently considered to include the four New World species, *A. filiculoides* Lamarck (type species), *A. caroliniana* Willdenow, *A. mexicana* Presl, and *A. microphylla* Kaulfuss, and (2) the *Rhizosperma* (nine floats), which is currently considered to include the two Old World species, *A. pinnata* R. Brown and *A. nilotica* DeCaisne. The two Old World species are readily distinguished in that *A. pinnata* exhibits a triangular shape and *A. nilotica* is a gigantic species. The four New World species may exhibit quite similar vegetative growth characteristics and collected material is often sterile, precluding the use of the major taxonomic criterion. Even when reproductive structures are present there can be confusion (cf. Ott & Petrik-Ott, 1973). The identification of *Azolla* species, especially in the section *Euazolla*, is often rather tenuous and their taxonomy appears to be subject to revision (Peters et al., 1980b).

Biogeography

Azolla species are found in fresh-water ecosystems of temperate and tropical regions throughout the world. Since wind and wave action as well as other turbulence causes fragmentation and diminished growth, *Azolla* is not found on large lakes or swiftly moving waters (Ashton, 1974). It is, however, capable of luxurious growth in more placid environments such as ponds, marshes, canals, drainage ditches and, significantly, rice paddies. Although *Azolla* can colonize bodies of water that are nitrogen deficient, enabling them to grow where other aquatic plants cannot, their growth can be limited by the availability of other nutrients, especially phosphorus and iron (Olsen, 1972; Talley et al., 1977; Watanabe et al., 1977; Watanabe, 1978; Singh, 1979).

Several studies have considered the geographical distribution of one or more *Azolla* species (Svenson, 1944; Schulthorpe, 1967; Moore, 1969; Sweet & Hills, 1971). Lumpkin and Plucknett (1980) recently presented a comprehensive literature survey of the geographical distribution of the extant *Azolla* species, along with a map of their world distribution. Prior to marked human influences, *A. caroliniana* was restricted to the eastern

United States and Caribbean; *A. filiculoides* occurred in southern South America northward into western North America, including Alaska; *A. mexicana* was found in northern South America through western North America to British Columbia, and east to Illinois; and *A. microphylla* was indigenous to western and northern South America, subtropical North America, and the West Indies. *A. pinnata* was present in much of the Eastern Hemisphere, that is, much of Asia and the coast of tropical Africa, and *A. nilotica* from the upper reaches of the Nile to the Sudan.

As illustrated in Lumpkin and Plucknett (1980), man has significantly altered the original species distribution. *A. filiculoides* is now found in Europe, where it may have been indigenous prior to the last Ice Age (Sculthorpe, 1967), Asia and Australia, and *A. caroliniana* has been reported to occur in Asia, South America, and Europe. Man appears to have had less effect on the further distribution of *A. mexicana, A. microphylla*, and *A. pinnata*, and virtually no effect on the distribution of *A. nilotica*. It should be borne in mind that there are necessarily questions in regard to whether some identifications in the literature are correct for the *Euazolla* species, since they are often based on nonsporulating material (cf. Ott & Petrik-Ott, 1973 and Pieterse et al., 1977).

General morphology and anatomy

Sporophytes of *A. caroliniana, A. filiculoides, A. mexicana*, and *A. pinnata* are shown in Fig. 4a–d. The sporophytes consist of multibranched, prostrate, floating stems (rhizomes) that bear deeply bilobed leaves and determinant, adventitious roots. The extensive branching pattern results in numerous stem apices and a growth habit that ranges from flabellate to polygonal, depending upon the degree and pattern of fragmentation. An abscission layer at the point of root and branch attachment facilitates vegetative propagation through fragmentation (Rao, 1936). The diameter of the sporophyte is usually 1–3 cm, with *A. pinnata* being the smallest and *A. filiculoides* the largest. Larger plants of each species can occur under conditions that limit fragmentation. The following anatomical description is based on the studies of Konar and Kapoor (1972) and Gunning et al. (1978) for *A. pinnata*; Campbell (1893, 1895), Smith (1955), and Bonnet (1957) for *A. filiculoides*; and our own observations on these species as well as *A. mexicana* and especially *A. caroliniana*.

The sporophyte has a dorsiventral organization. Each rhizome apex has an upcurved meristem supporting a three-sided apical cell (Fig. 5). Its two cutting faces produce daughter cells in two lateral ranks. Anticlinal division of each daughter results in a four-celled transverse profile with the cells arranged like quadrants of a circle. Roots and lateral branches are derived from further formative divisions of the ventral cells whereas leaves are derived from the dorsal cells.

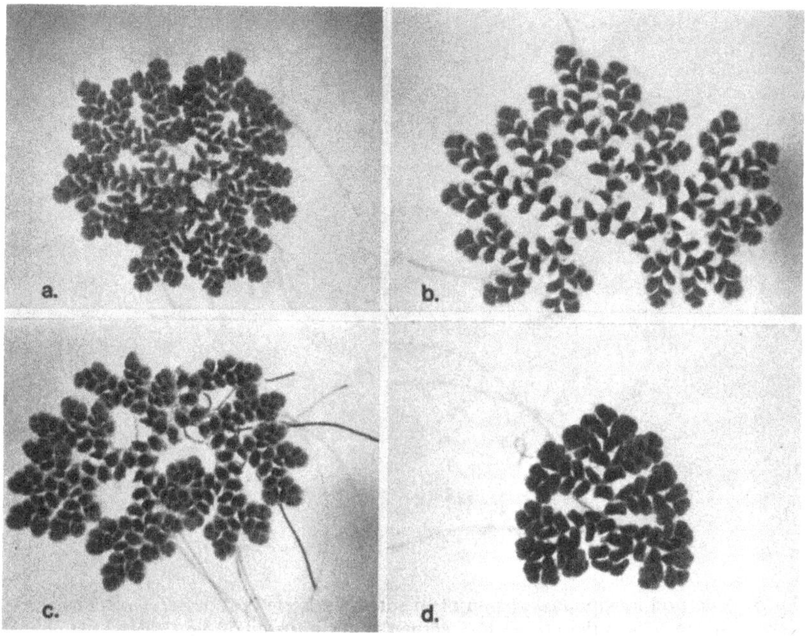

Fig. 4. Habit micrographs of *A. caroliniana* (a), *A. mexicana* (b), *A. filiculoides* (c), and *A. pinnata* (d). Each dorsal leaf lobe is approximately 1 mm long in these species.

The mature rhizome shows the relatively simple anatomy characteristic of ferns. The vascular bundle is protostelic in organization. Phloem forms bicollaterally on the dorsal and ventral sides of the xylem. The protoxylem consists of only one or two elements and is dorsal to the metaxylem, separated from it by a layer of xylem parenchyma. The metaxylem usually comprises two spiral vessel elements. A single layer of pericycle surrounds the vascular tissue followed by an endodermis with Casparian thickenings in the lateral anticlinal cell walls. The cortex is only a few cells thick. It is composed of parenchymatous cells, which are chlorophyllous and tightly packed with minimal intercellular space. The rhizome is covered by a simple single-layered epidermis.

Adventitious roots occur on the ventral side of the rhizome at most nodes, hanging freely into the water. Mature portions of the root are covered with root hairs. The apex and young portion of each root are covered by a sheath and two-layered cap. Elongation frees the root from the sheath and cap, and hairs emerge from epidermal cells. The anatomical development of the root is well ordered. The apical cell has a four-sided pyramidal shape with its base oriented toward the direction of root-growth. All four sides are cutting faces. The basal face contributes cells to the root sheath and cap, whereas the three lateral faces contribute

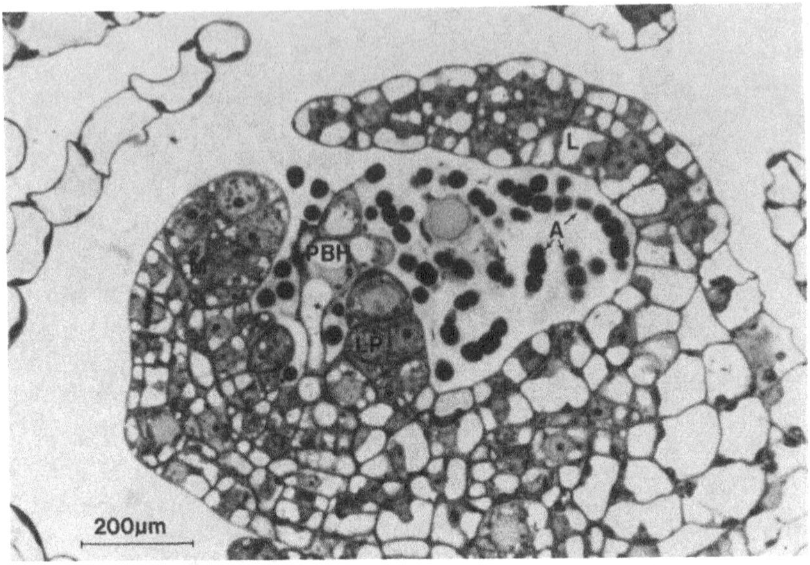

Fig. 5. A vertical longitudinal section of an apical meristem of *A. caroliniana*. The darkly staining cells in the pocket formed by the upcurved meristem (M) and the young overarching dorsal leaf lobe (L) are vegetative *Anabaena* cells (A), which comprise the generative apical colony. Note the young primary branched hair (PBH) emerging from the axil of a leaf primordium (LP).

equally to the remainder of the root. The mature root has a simple epidermis and a two-layered cortex. The outer cortex is composed of a single layer of 9 or 12 large parenchymatous cells, whereas the inner cortical layer consists of six smaller cells. Inside the cortex there are two more concentric rings of cells. The outer ring is formed by six endodermal cells and the inner ring is composed of six pericycle cells. The latter surround the vascular tissue, which is a simple bicollateral protostele.

Two lateral rows of leaves are borne alternately on the rhizome and may overlap to such an extent that it is obscured. Each leaf has two lobes of approximately equal size, one dorsal and the other ventral. The thin ventral lobe is nearly colorless and floats on the water surface. The distal half of its blade is only one cell thick. The proximal half, however, has a recognizable adaxial and abaxial epidermis and, near the rhizome, a limited spongy mesophyll with intercellular spaces.

The mature dorsal leaf lobe is aerial and has a clearly defined multilayered mesophyll as well as adaxial and abaxial epidermal tissues. The abaxial epidermis has many stomata. Each stoma initially differentiates a pair of guard cells that fuse their cytoplasms into a single annular guard cell with a central stomatal pore at maturity. Single-celled papillae are

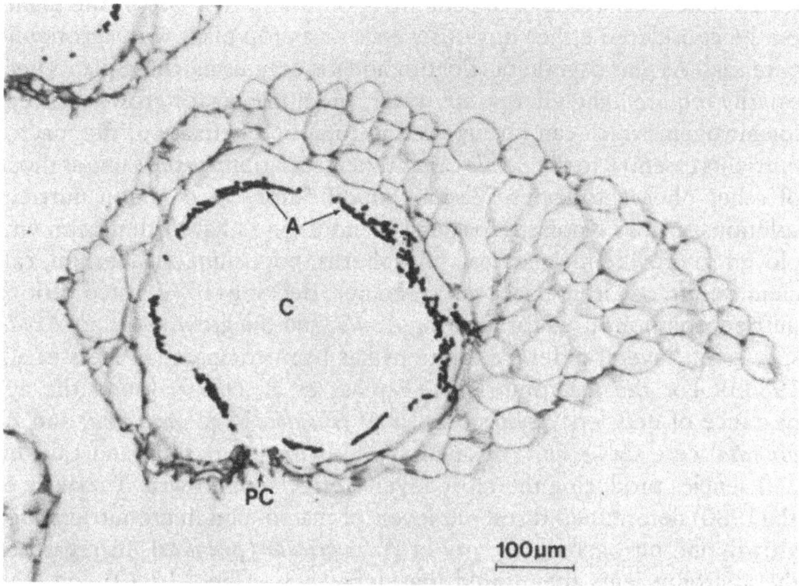

Fig. 6. An oblique longitudinal section of a dorsal leaf lobe of *A. caroliniana*. The large cavity (C) containing *Anabaena* filaments (A) is conspicuous in the proximal half of the lamina. Note the cavity pore closure (PC) along the cavity wall.

also present on the abaxial epidermis. These differentiate from epidermal cells, which form an outward protuberance that is cut off by a periclinal crosswall. The outer cell enlarges and becomes turgid, forming the papillae. They vary somewhat in shape and distribution among species. On *A. filiculoides* and *A. caroliniana* the papillae appear randomly distributed and have an isodiametric shape. In *A. pinnata* and *A. mexicana* they are more oblong and are ordered in rows along the leaf surface. The leaf margins are entire in all species. Each mature dorsal leaf lobe has an ellipsoid cavity in the proximal half of its lamina (Fig. 6). The lining of this extracellular space is formed by an infolding of adaxial epidermis. The cavity normally contains the endophyte, *Anabaena azollae*. However, endophyte-free plants may rarely occur in nature and can be generated by several techniques (Peters & Mayne, 1974a; Hill, 1975, 1977).

Growth and mineral nutrition

Azolla species are capable of very rapid vegetative growth, with *A. caroliniana, A. filiculoides, A. mexicana*, and *A. pinnata* capable of doubling their biomass in 2 days or less under optimal laboratory conditions (Peters et al., 1980b). Under natural conditions this rapid vegetative propagation enables these organisms to completely cover moderately protected water surfaces and, depending upon where they occur, the plants can be considered either noxious weeds or a crop plant with agronomic potential. As photosynthetic, floating aquatic organisms, these plants necessarily require light, energy, air, water, and nutrients for growth. Except for nitrogen, which can be supplied in total by N_2 fixation, the macronutrients essential to the *Azolla-Anabaena* association are the same as those of other photoautotrophs. Various modifications of common nutrient solutions such as Knop's, Hoagland's, and Crone's have often been employed to provide the required phosphorus, potassium, magnesium, calcium, sulfur, and iron in laboratory studies. Becking (1979) listed various nutrient solutions used for growing *Azolla*, and the growth of four *Azolla* species on several different solutions has been compared (Peters et al., 1980b). For the macronutrients, Espinas et al. (1979) found the appearance of deficiency symtpoms in *A. filiculoides, A. mexicana*, and *A. pinnata* were Ca > Fe > P > Mg > K, with magnesium and calcium deficiencies producing the most adverse effects on growth. Yatazawa et al. (1980) determined threshold levels of macro- and micronutrients for growth and nitrogenase activity of *A. imbricata* (*pinnata*). In regard to the micronutrients they found that deficiency of Fe, Mg, Co, Zn, Cu, Mo, or B had unfavorable effects on growth and N_2 fixation. Earlier studies had shown that cobalt (Johnson et al., 1966) and molybdenum (Bortels, 1940) were essential micronutrients for the symbiotic association grown in the absence of combined nitrogen sources. Elevated requirements for Fe, Mo, and Co are common in organisms that contain nitrogenase.

The flat spray growth mode enables the common *Azolla* species to achieve a high plant density with the effective utilization of incident solar energy for photosynthesis and associated processes. Although normally free floating, the roots of *Azolla* are capable of penetrating into water-laden soils and plants may become "rooted" in shallow water. It is unresolved as to whether uptake of nutrients occurs only via the roots or through the floating stem and ventral leaf lobes as well. Although laboratory studies suggest that uptake may occur in plants bearing very few roots, rooted plants growing adjacent to floating plants in a natural environment often appear green and healthy, whereas the floating plants exhibit signs of phosphate deficiency. Furthermore, endophyte-free plants and plants purposely stressed for specific nutrients often exhibit increased root production.

There are reports of both adverse and beneficial effects of combined nitrogen on *Azolla* growth and N_2 fixation (cf. Lumpkin & Plucknett, 1980). In the authors' experience the effect of combined nitrogen sources will be determined by a number of factors. These include concentration and pH effects, duration of culture on the combined nitrogen source, and whether or not the plants had epiphytic contaminants. Most adverse effects can be attributed to stimulated growth and increased competition from other organisms. Any studies pertinent to the effect of combined nitrogen on *Azolla* itself must be conducted with cultures freed of epiphytic organisms. In such cultures low concentrations of a specific combined nitrogen source can stimulate growth and not significantly affect N_2 fixation. However, increasing concentrations of combined nitrogen result in a decline of nitrogenase activity and at high concentrations, for example, above 25 m*M* nitrate, a diminished growth rate as well (Peters et al., 1981a).

Rates of growth and the contributions to total nitrogen input from N_2 fixation and various concentrations of ammonium, nitrate, and urea have been determined by employing ^{15}N-labeled combined nitrogen sources and isotope dilution techniques (Peters et al., 1981a). The doubling times for the *A. caroliniana* association clustered around 2 days for controls and cultures provided with combined nitrogen sources. In general, although there was a decrease in nitrogenase activity with increasing combined nitrogen, the nitrogen content of the plant material remained quite constant. The isotope dilution studies showed that this was due to the concurrent utilization of the combined nitrogen source and the nitrogen from N_2 fixation.

Composition

A dry weight from 4.8% to 7.7% of the fresh weight has been reported for *Azolla* species (Moore, 1969; Buckingham et al., 1978; Peters et al., 1980b), and numerous unpublished determinations of the four species in our studies have yielded a value of 6.02% ± 1.02%. Although the dry matter usually contains 3%–6% nitrogen (Moore, 1969; Talley et al., 1977; Watanabe et al., 1977; Peters et al., 1980b; Talley & Rains, 1980b), the nitrogen content is strongly influenced by growth conditions. For example, Talley and Rains (1980b) reported 30 values ranging from 2.2% to 5.6% as a function of varying light intensities and day/night temperature regimes. Under optimized growth conditions nitrogen content is consistently in the range of 5%–6.5%, and carbon accounts for 40%–43% of the dry weight (Peters et al., 1980b, 1981a, b). In an analysis of the nutritive value of *A. filiculoides* for rats, Buckingham et al. (1978) found the dry matter was composed of approximately 15.5% ash, 26.6% acid detergent fiber, 39.2% neutral detergent fiber, 15.2% cellulose, 9.3% lignin, 4.5% nitrogen, and 5.0% fats (ether extract). *Azolla* had low nutritive value as the sole protein source for rats, and deficiencies in lysine, methionine, and histidine were indicated. However, in accord with our unpublished observations for *A. caroliniana*, the amino acid composition of *A. filiculoides* compares quite well with that of good-quality alfalfa. The low nutritive value of *Azolla* for rats is at

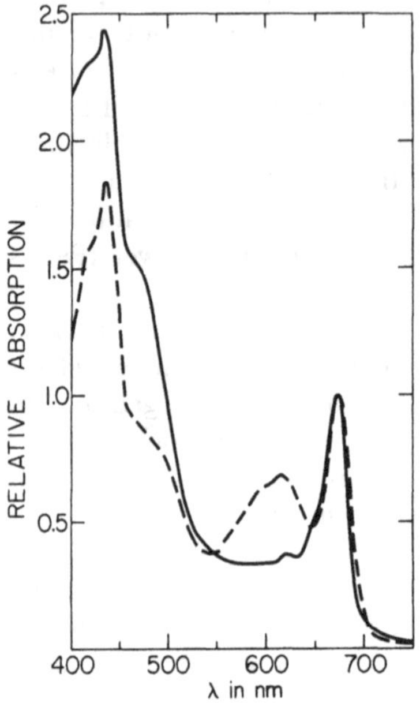

Fig. 7. Absorption spectra of cell free extracts from the endophyte-free *Azolla* (—) and the endophytic *Anabaena* (---). The spectra have been normalized at 673 nm. (Reproduced from Ray et al., 1979, with permission.)

least partially due to the high neutral detergent fiber, high mineral content, and low digestibility of *Azolla* nitrogen (Buckingham et al., 1978).

Whole-plant physiology and biochemistry

Photosynthesis

Since *Azolla* chloroplasts contain chlorophylls *a* and *b* as well as carotenoids, while the *Anabaena* filaments contain chlorophyll *a*, phycobiliproteins, and carotenoids, the light-harvesting pigments of the partners are complementary (Fig. 7). In the *A. caroliniana-Anabaena* association the endophyte accounts for 10%–20% of the association's total chlorophyll and about 16% of its total protein with phycobiliproteins accounting for 4%–10% of the endophyte's protein (Peters & Mayne, 1974a; Peters, 1978; Ray et al., 1978). Phycocyanin (λ max 610 nm) accounts for about 70% of the endophyte's phycobiliprotein, whereas phycoerythrocyanin (λ max 570 nm, shoulder 590 nm) and allophycocyanin (λ max 647 nm, shoulder 620 nm) account for about 17% and 13%, respectively (Tyagi et al., 1980). In addition to photosynthetic pigments, *Azolla* may also contain anthocyanins, predominately luteolinin-5-glucoside, with lesser amounts of apigeninidin glucoside (Holst, 1977; Pieterse et al., 1977). Anthocyanin formation can be triggered by a variety of environmental factors. Their production can cause the *Azolla* to take on a variety of reddish hues.

The method of preference for obtaining sufficient quantities of the endophyte for most physiological and biochemical studies yields filaments of the endophyte from all stages of leaf development and simply involves rupturing of the leaf cavities by applying mild pressure with a roller, filtration to remove plant debris, and slow-speed centrifugation to pellet the *Anabaena* filaments (Peters & Mayne, 1974a). The endophyte can also be isolated from individual leaf cavities as packets surrounded by a thin limiting envelope using cellulytic enzymes (Peters, 1976; Peters et al., 1978) or by removing the filaments from individual leaves under the dissection microscope (Peters, 1978; Peters et al., 1980a). Moreover, *Azolla* chloroplasts and *Anabaena* filaments have been separated on sucrose density gradients and characterized with respect to partial reactions of photosynthesis, delayed light emission, and P700 content (Peters & Mayne, 1974a).

Photosynthesis in the *A. caroliniana-Anabaena* association, the endophyte-free *Azolla*, and the endophytic *Anabaena* has been characterized by determining action spectra, intermediates of CO_2 fixation, CO_2 compensation points, and the effect of O_2 partial pressures (Peters et al., 1979; Ray et al., 1979). As with other cyanobacteria, the relative quantum yield for photosynthesis in the action spectrum of the endophyte is highest between 580 and 640 nm, the region of phycobilin absorption. However, there is no readily detectable contribution by the endophyte to the association's action spectrum. Action spectra for photosynthesis in the association and endophyte-free *Azolla* are very similar to one another and to other green plants, with the maximum quantum yield occurring between 650 and 670 nm. The association and individual partners exhibit Calvin cycle (C3) intermediates of CO_2 fixation (Ray et al., 1979). Sucrose is a primary fixation product in the *Azolla* but does not occur as a ^{14}C-labeled reaction product in the endophyte. In the association, CO_2 fixation and growth (Peters et al., 1980b) are saturated at a photosynthetic photon flux density (PPFD) of about 400 $\mu E \cdot m^{-2} \cdot s^{-1}$. As with other C3 plants, the association and endophyte-free *Azolla* exhibit an O_2-dependent, CO_2 compensation point and photosynthesis is inhibited by atmospheric O_2. Rates of CO_2 fixation in air are about 40% less than those at 2% O_2 and the aerobic CO_2 compensation point is about 40 ppm CO_2. Photosynthesis by the endophytic *Anabaena* is not inhibited by atmospheric O_2 and the CO_2 compensation point is about 4 ppm CO_2 at both 20% and 2% O_2. Although studies with the endophytic *Anabaena* isolated from the leaf cavities demonstrate its physiological capabilities, such studies do not, necessarily, extrapolate to the symbiotic state. The endophyte is known to exhibit morphological and physiological changes as a function of leaf ontogeny, and it is probable that the microenvironment of the leaf cavity also changes during leaf ontogeny. Recently, however, it has been estimated that the endophytic

Anabaena contributes 6%–10% of the total photosynthetic capability of the association (Kaplan & Peters, unpublished), and current studies suggest that there is an interaction in the carbon metabolism of the host and endophyte (see the section dealing with developmental physiology).

Nitrogen fixation

The reduction of N_2 to ammonia (fixed nitrogen) is catalyzed by the enzyme nitrogenase. This enzyme is made up of two different proteins, the molybdenum iron protein (MoFe protein) and the iron protein (Fe protein), both of which are O_2 labile. Thus, the enzyme must be protected from both atmospheric O_2 and the intracellular O_2 produced by photosynthesis. As with free-living heterocystous cyanobacteria, the endophytic *Anabaena* is able to reduce N_2 under an air atmosphere and the nitrogenase is assumed to be localized in the heterocysts. The absence of CO_2 fixation and the rapid reduction of triphenyltetrazolium chloride (TTC) to red formazan in heterocysts of the endophytic *Anabaena* (Peters, 1975) support this concept.

Nitrogenase activity requires a source of ATP and reductant. While N_2 is the natural substrate, nitrogenase is capable of reducing a number of other substrates, the most notable being the reduction of acetylene to ethylene and the ATP-dependent reduction of protons to H_2 in the absence of any other reducible substrate. Although the reduction of N_2 to 2 NH_3 requires the transfer of six electrons, the reduction of C_2H_2 to C_2H_4, or protons to H_2, requires two electrons. The acetylene reduction assay provides a very sensitive and economical means of estimating nitrogenase activity and the theoretical relationship of 3 C_2H_2 reduced/N_2 reduced is often used to extrapolate from C_2H_2 reduction to N_2 fixation. However, the actual relationship in in vivo studies is quite variable. It is subject to many factors and should be determined experimentally for each system.

In the *A. caroliniana-A. azollae* association and in the isolated *Anabaena*, C_2H_2 reduction, $^{15}N_2$ fixation, and ATP-dependent H_2 production have been used to study nitrogenase activity. Under anaerobic dark conditions the reduction of all substrates is negligible. Dark aerobic reductions occur but they are dependent upon the endogenous supplies of reductant accumulated during prior photosynthesis, and rates are almost always less than 40% of those obtained aerobically in the light. The reduction of all substrates is maximal under anaerobic or microaerobic conditions in the light. Figure 8 shows the effects of O_2, darkness, and light on nitrogenase-catalyzed C_2H_2 reduction.

At a saturating concentration of acetylene (0.1–0.15 atm) there is little or no detectable nitrogenase-catalyzed H_2 production (Peters et al., 1976, 1977, 1981a). However, the amount of H_2 that can be measured

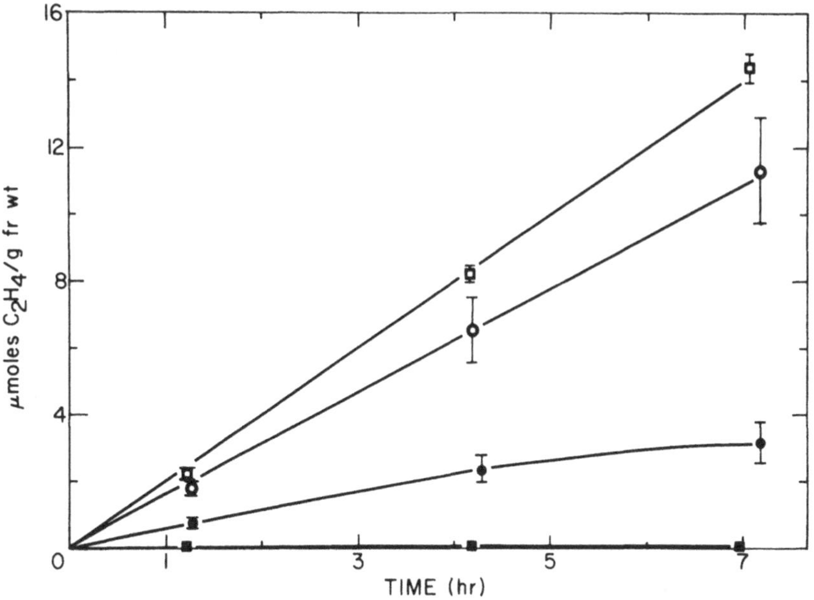

Fig. 8. Nitrogenase-catalyzed reduction of acetylene (C_2H_2) to ethylene (C_2H_4) in the *A. caroliniana-A. azollae* association using anaerobic-light (□), aerobic-light (○), aerobic-dark (●), and anaerobic-dark (■) assay conditions. Each data point is the mean ± standard deviation of triplicates.

under argon or various N_2 partial pressures depends upon the amount of unidirectional hydrogenase activity present in the endophyte (Peters, 1977; Peters et al., 1977, 1981a). Since this enzyme oxidizes the H_2 produced by nitrogenase, recycling electrons and ATP into the system, rates of H_2 production under argon are always less than rates of acetylene reduction (Peters et al., 1976, 1977). Although the occurrence of unidirectional hydrogenase also complicates determinations of the inhibition of H_2 production by N_2, H_2 production is markedly inhibited as the N_2 concentration is increased (Peters et al., 1977). The C_2H_2/N_2 conversion factor for the *Azolla-Anabaena* association and isolated endophyte has been determined under a variety of experimental conditions (Peters et al., 1977, 1980a, 1981b; Kaplan & Peters, 1981). By monitoring C_2H_2 reduction throughout the light and dark intervals of a 16 h–8 h light-dark photoperiod and obtaining conversion factors at the midpoint of the light and dark periods it was recently estimated that during a 24-h interval, 81% of the nitrogen input from N_2 fixation occurred during the light period (Peters et al., 1981b).

Relationship between photosynthesis and N_2 *fixation*

Photosynthesis is the ultimate source of all reductant and ATP for nitrogenase activity in these associations. The dependence of dark aerobic nitrogenase activity upon endogenous reserves of photosynthate can be demonstrated by varying the length of light or dark periods to maintain or deplete the endogenous reserves. Diminished rates under aerobic dark conditions versus those obtained under aerobic light conditions with DCMU (Peters, 1975, 1976) imply that dark, respiratory-driven nitrogenase activity may be ATP limited. If one assumes that reductant is generated in the same manner in the light and dark, then the increased activities in the light are necessarily attributed to photosynthetically generated ATP. Although photosystem II activity is required to provide photosynthate via CO_2 fixation, it is not directly required for nitrogenase activity. This has been demonstrated by simultaneous measurements of CO_2 fixation and nitrogenase-catalyzed C_2H_2 reduction on the same sample (Peters et al., 1980a). The cumulative results of a number of studies (Peters & Mayne, 1974b; Peters, 1975, 1976; Peters et al., 1980a) strongly imply that reductant provided by prior photosynthesis and cyclic photophosphorylation are the primary driving forces of nitrogenase activity in the light. The interaction of photosynthesis with N_2 fixation has also been demonstrated by obtaining the action spectra for nitrogenase-catalyzed acetylene reduction in the association and in isolated *Anabaena* (Tyagi et al., 1981). Although phycobiliproteins are generally considered to be accessory pigments for photosystem II and to be depleted or absent in heterocysts, heterocysts of the endophytic *Anabaena* were found to retain phycobiliproteins. Moreover, these pigments, which absorb light energy in that portion of the visible spectrum least effectively used by the *Azolla* pigments, were found to effectively capture light energy for use in driving nitrogenase-catalyzed acetylene reduction.

Transfer of fixed nitrogen from Anabaena *to* Azolla

In the absence of a combined nitrogen source, N_2 fixed by the endophytic *Anabaena* provides the total nitrogen requirement of the association. The endophyte in mature leaf cavities releases newly fixed N_2 as ammonia, which is rapidly assimilated by the *Azolla*. These conclusions are based on the following experimental results. When exposed to $^{15}N_2$ the endophytic *Anabaena* releases up to 50% of the N_2 it fixes into the incubation medium as ammonium (Peters, 1977; Peters et al., 1980a). Since it is probable that some of the released ammonium was reassimilated by undifferentiated filaments in these preparations, it is equally probable that mature filaments release more than half the N_2 they fix. Related studies, which compare supernatants after incubating the endophyte under Ar-$^{14}CO_2$ and N_2-$^{14}CO_2$ (Ray et al., 1978; Peters et al., 1980a) indicated that there was little or no release of organic nitrogen compounds such as amino acids.

The release of newly fixed nitrogen as ammonia by the isolated endophyte and the occurrence of free ammonia in the tissues and/or leaf cavities of the N_2-grown association led to studies of the enzymes known to be involved in ammonia assimilation. Both the association and endophytic *Anabaena* exhibit glutamine synthetase (GS), glutamate synthase (GOGAT), and glutamate dehydrogenase (GDH) activities. Although both partners must be considered capable of assimilating ammonia, the *Azolla* was estimated to account for about 90% of the association's total GS activity and 80% of its total GDH activity (Ray et al., 1978). These values reflect an average of the activities and distribution of these enzymes in all developmental stages of both partners. Incubation of the association under $^{15}N_2$-enriched air followed by chase periods with air showed a low percentage of the total ^{15}N in the ammonia fraction with a rapid incorporation first into ethanol-soluble compounds (amino acids, etc.) and then into ethanol-insoluble compounds (proteins, etc.) (Peters et al., 1979). The ammonia-assimilating enzymes and their partitioning between the partners as a function of leaf development are considered next in conjunction with their developmental physiology.

Development

Growth and development of the host and the symbiont are synchronous (Hill, 1977). Each apical meristem is supported above the water surface by its stem axis and has a small colony of *Anabaena* filaments associated with it. The colony is situated in a pocket formed against the upcurved meristem by the young leaves (Fig. 5). Filaments of the apical colonies are composed entirely of small vegetative cells. The size of the apical colonies is very consistent from apex to apex or from plant to plant. Even when *Azolla* plants growing at different rates are compared, the apical *Anabaena* cell mass is similar in size. One never sees, for example, *Anabaena* overgrowing the apices of a slow-growing fern, nor is it possible to deplete the apical colonies by accelerating the fern's growth rate. Thus the growth rate of the endophyte is coordinated with that of its host.

The establishment of *Anabaena* filaments in each leaf begins in the young leaves contiguous with the apical *Anabaena* colony. The process is complete by the time each leaf emerges as a distinct organ visible to the naked eye. Development of the cavity begins in the young leaves, which are rapidly enlarging by cellular expansion. An area of premesophyll tissue at the base of the dorsal leaf lobe enlarges very little compared to the more distal mesophyll, causing a depression to form in the adaxial epidermis (Fig. 9). As the apical meristem continues to grow and becomes further displaced from each leaf, filaments of the apical *Anabaena* colony become associated with each developing leaf in the area of its forming cavity.

The actual partitioning process appears to involve a specialized epi-

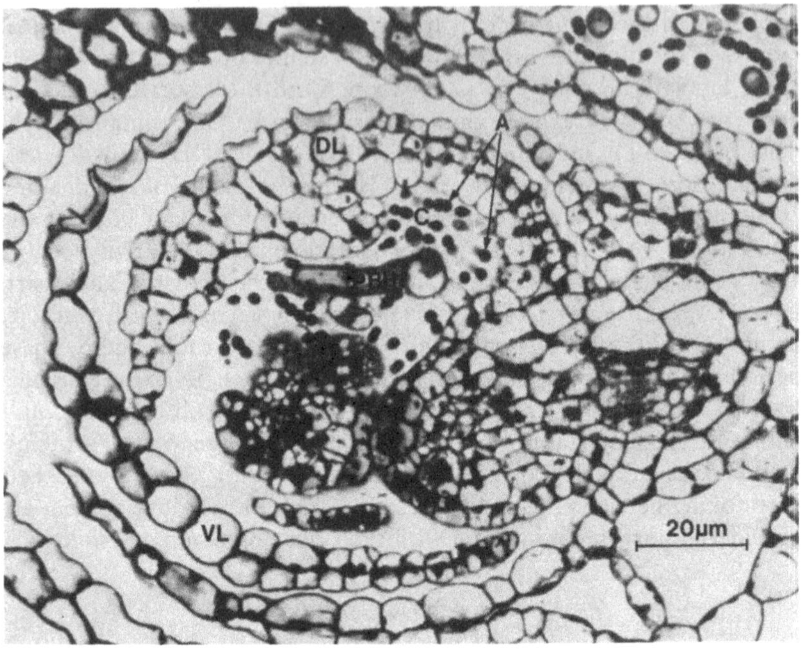

Fig. 9. A vertical longitudinal section of an apex of *A. caroliniana* showing the formation of the leaf cavity (C) in the base of a developing dorsal leaf lobe (DL). The primary branched hair (PBH) is located at the proximal edge of the depression. Ventral leaf lobe (VL), *Anabaena* cells (A).

dermal hair called the primary branched hair (PBH) (Calvert & Peters, 1981) (Fig. 10). Each leaf has only one primary branched hair. It originates from the axil of the leaf while the latter is still a primordium on the apical meristem and rapidly differentiates a single stalk cell supporting several elongate terminal cells (Fig. 5). Its growth is directed into the apical *Anabaena* colony where the terminal cells develop a network of ramified cell wall elaborations and the organelle-rich cytoplasm characteristic of transfer cells (Pate & Gunning, 1972). These cells have many ribosomes and mitochondria, abundant rough endoplasmic reticulum, and several proplastids (Fig. 11). Plasmodesmata are conspicuous on the crosswalls between each terminal cell and the stalk cell. These attributes suggest that the hairs may be metabolically interactive with the associated apical *Anabaena* filaments. Though there is no proof of interaction, the *Anabaena* filaments seem to be attracted to the PBHs in the apical region.

The PBHs of the youngest three or four leaves are all in contact with the apical *Anabaena* colony. As continued growth displaces the eldest of these leaves from the meristem, its PBH is dissociated from the apical colony (Fig. 12). Those *Anabaena* filaments entangled in the hair are

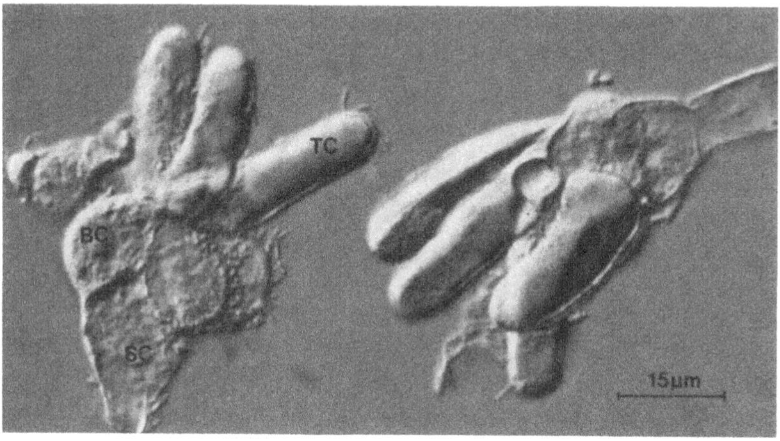

Fig. 10. The primary branched hair of the *Azolla* leaf cavity. The PBH is composed of a stalk cell (SC), two to three body cells (BC), and four to five terminal cells (TC). These isolated hairs were photomicrographed with differential interference contrast microscopy.

separated from the apical mass and remain associated with the hair. Having developed in the leaf axil, the PBH is now positioned at the proximal end of the cavity depression (Fig. 9). Cells around the rim of the depression become meristematic and produce epidermal cells, which form the cover of the cavity like the closing of an iris diaphragm. During this growth the PBH, with its associated *Anabaena* filaments, is engulfed by the forming cavity (Fig. 13). In our interpretation of the leaf inoculation mechanism, the PBH can be likened to an inoculation needle. It enters a stock culture (the apical *Anabaena* colony), becomes covered with the microorganism, and then delivers the microbe to a new culture container (the leaf cavity).

As the adaxial epidermis is forming the cover over the leaf cavity, some *Anabaena* filaments begin to differentiate heterocysts. Concurrently, many additional epidermal hairs begin to form on the cavity walls. One of these, the secondary branched hair (Fig. 14), is positioned at the back of the cavity depression (Calvert & Peters, 1981). The pattern of its development is similar to the PBH except that it differentiates fewer terminal cells. The other hairs, termed *simple hairs* (Calvert & Peters, 1981), also begin to emerge from the cavity walls at this time. These hairs are morphologically distinct from the branched hairs, being elongate, terete structures consisting of only two cigar- to barrel-shaped cells (Fig. 15). They constitute a population distinct from the branched hairs. The terminal cell of each simple hair quickly differentiates transfer cell ultrastructure, as does the distal portion of the stalk cell (Fig. 16). The number of simple hairs increases during the isolation and enclosure of the *Anabaena* filaments into each leaf, and the branched hairs undergo

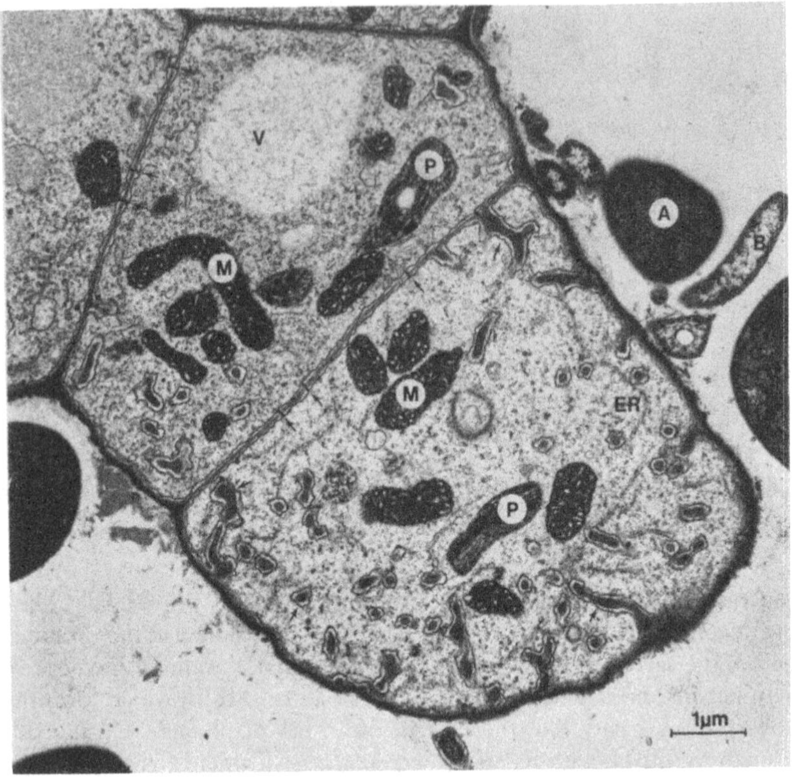

Fig. 11. A thin section of a portion of a primary branched hair in a leaf cavity of *A. caroliniana*. The cells have an organelle-rich cytoplasm and numerous cell wall elaborations (arrows) characteristic of transfer cell ultrastructure. Plasmodesmata are abundant in the crosswalls (arrowheads). Mitochondria (M), endoplasmic reticulum (ER), plastid (P), vacuole (V), *Anabaena* (A), bacteria (B).

further differentiation. Two or three isodiametric cells are cut off the distal end of the PBH's stalk cell, intercalary to the terminal cells (Fig. 10, 14). These cells differentiate transfer cell ultrastructure as the terminal cells senesce (Calvert & Peters, unpublished observation).

As leaf development continues it becomes clear that the branched hairs are located in similar positions in every leaf cavity, always lying along the path of the dorsal lobe foliar trace. The trace is two to three cell layers below the epidermis, precluding direct contact with the branched hairs. Transfer cell ultrastructure is present in the foliar trace in the region of each branched hair (Fig. 17). It is best developed in xylem parenchyma, where most wall elaborations form on the walls contiguous with the xylem vessels. Some wall elaborations can also be found in phloem parenchyma and in those cells separating the leaf trace from that part of the cavity wall bearing the branched hair.

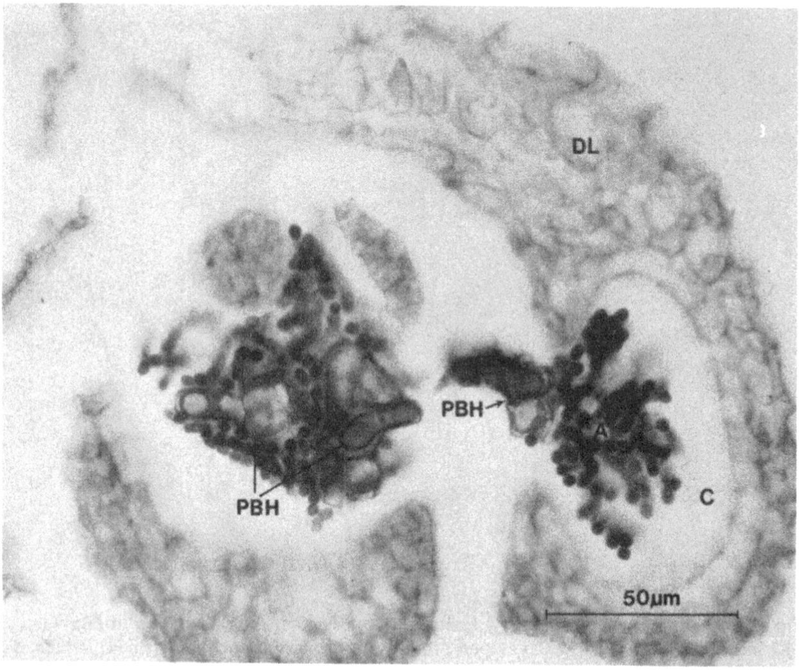

Fig. 12. A 7-mm-thick horizontal longitudinal section of paraffin-embedded tissue from an apical region of *A. caroliniana*. The dark mass on the left is the apical *Anabaena* colony. Portions of two primary branched hairs (PBH) are visible in the mass of endophyte cells. On the right is a developing dorsal leaf lobe (DL) with forming cavity (C). A mass of *Anabaena* has been separated from the apical colony by the young leaf's PBH.

As each leaf matures, the *Anabaena* filaments in its leaf cavity multiply. In a mature leaf cavity the vegetative cells of the endophyte have increased in size and are filled with photosynthetic lamellae. The heterocyst frequency is 25%–30% and the filaments are concentrated along the walls of the cavity that have subjacent mesophyll. The 20–25 simple hairs in the mature cavity have developed transfer cell ultrastructure in their terminal cells. The branched hairs remain prominent along the foliar trace. While their body cells have differentiated transfer cell ultrastructure, their terminal cells have senesced.

Developmental physiology

Azolla-Anabaena associations are distinct from other plant-cyanobacteria symbioses (Stewart, 1977) in the synchronous development of the partners (Hill, 1975, 1977; Peters et al., 1980a; Calvert & Peters, 1981). The generative *Anabaena* filaments associated with the plant apex lack heterocysts and do not exhibit nitrogenase activity. As leaf cavities are

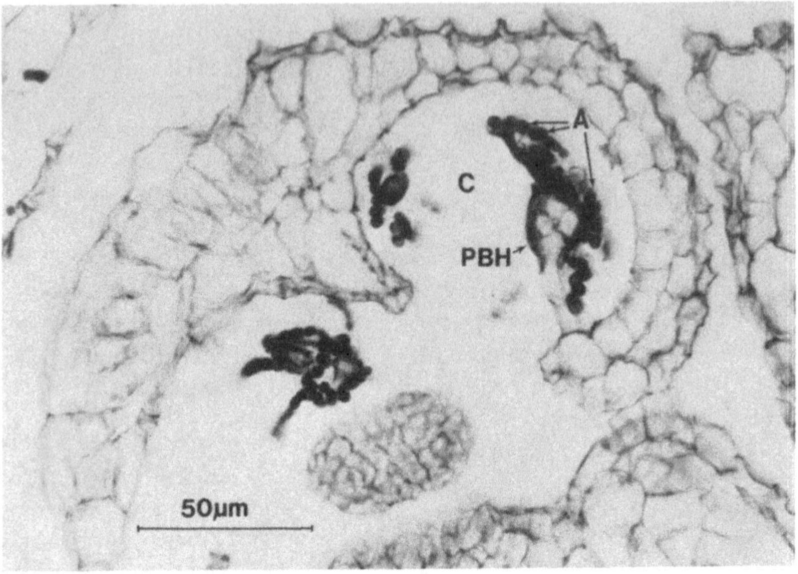

Fig. 13. A longitudinal section of a leaf slightly older than the one shown in Fig. 12. The leaf cavity (C) has now developed around the primary branch hair (PBH), which is surrounded by *Anabaena* filaments (A).

occupied by the endophyte, the rapid differentiation of heterocysts is accompanied by a rapid increase in nitrogenase activity. This developmental profile was originally described by Hill (1975) using *A. filiculoides*. It subsequently has been confirmed in another population of *A. filiculoides* as well as in *A. pinnata* (Shi et al., 1981) and studied in considerable detail in *A. caroliniana* (Peters et al., 1980a; Calvert & Peters, 1981; Kaplan & Peters, 1981). Physiological and biochemical studies of the whole association, endophyte-free plants, and populations of the endophyte isolated from all stages of development necessarily reflect a composite of activities and/or processes. The study of main stem axes, and individual leaves or segments of the axis bearing sequential groups of leaves, provides a more refined approach to an understanding of structure-function relationships and host-symbiont interactions.

The absence of nitrogenase activity, as determined with C_2H_2 reduction, in *Anabaena* filaments associated with the plant apex implied that N_2 fixed by the *Anabaena* in mature leaf cavities was transported to the apical region, meeting the nitrogen requirements of both the plant tissues and the generative *Anabaena* filaments. That fixed nitrogen is in fact transported, demonstrating interleaf interaction, has recently been demonstrated in main stem axes using a pulse-chase approach with $^{15}N_2$ (Kaplan & Peters, 1981). Although this approach enabled a concise dem-

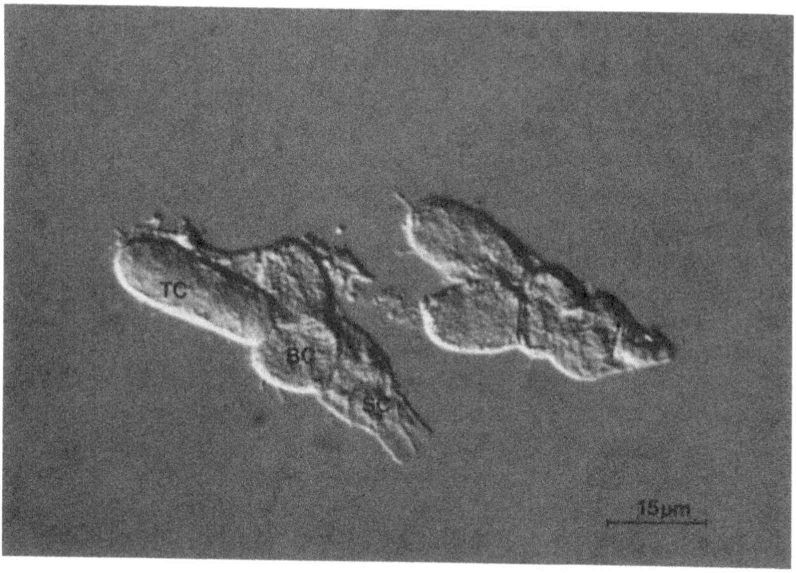

Fig. 14. Secondary branched hairs isolated from leaf cavities of *A. caroliniana*. These develop similarly to the PBHs but have fewer cells. Terminal cell (TC), body cell (BC), stalk cell (SC).

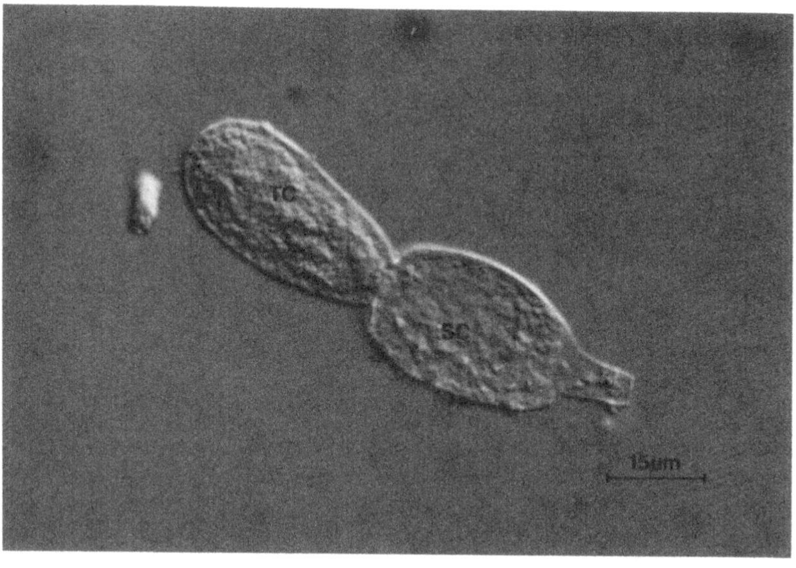

Fig. 15. An example of a simple hair isolated from an *A. caroliniana* leaf cavity. Terminal cell (TC), stalk cell (SC).

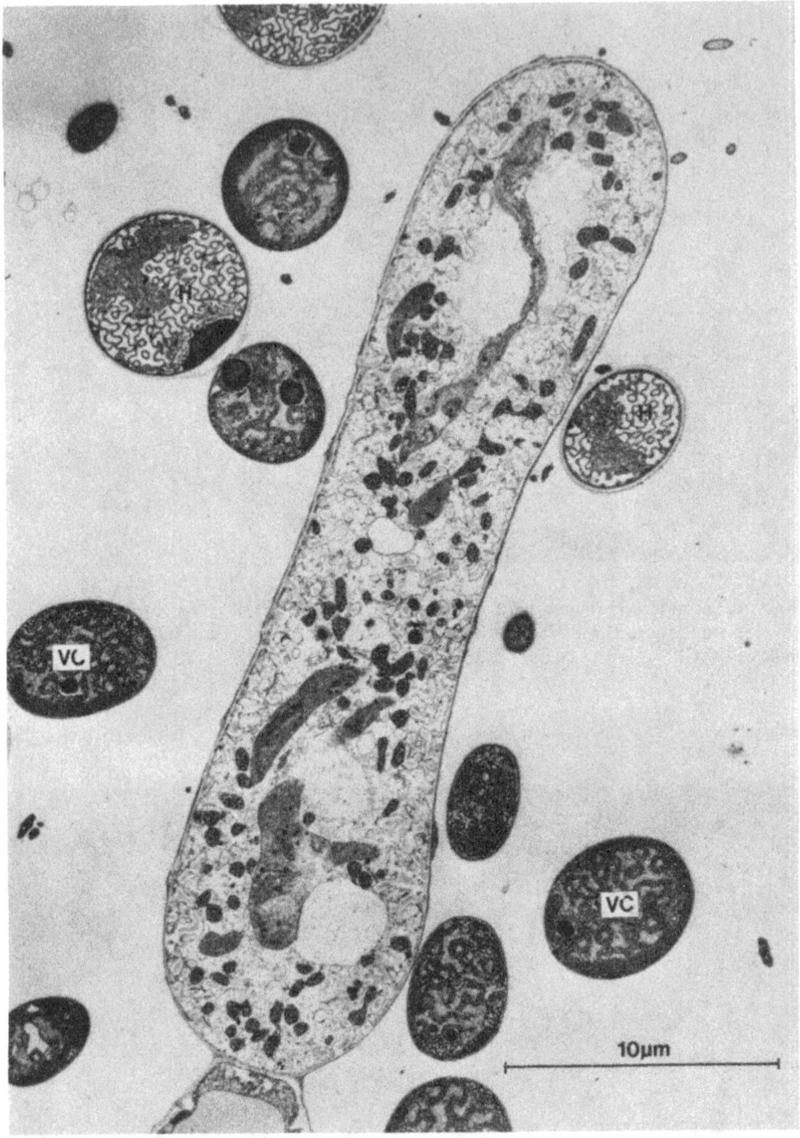

Fig. 16. Longitudinal thin section of the distal portion of the stalk of cell of a simple hair from *A. caroliniana*, showing the typical transfer cell ultrastructure that develops in these hairs. Note the *Anabaena* vegetative cells (VC) and heterocysts (H) adjacent to the hair.

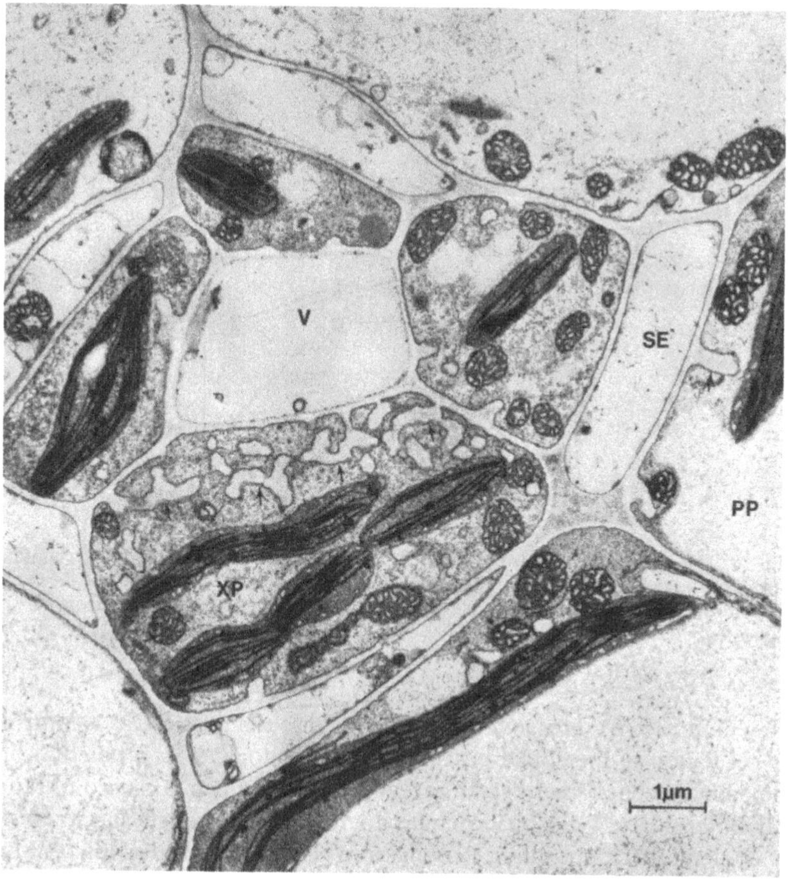

Fig. 17. The vascular trace of the dorsal leaf lobe in the region of the primary branched hair. Note the transfer cell differentiation in the xylem parenchyma (XP) with most wall elaborations (arrowheads) oriented toward the vessel (V). Wall elaborations (arrowhead) are also observed but to a lesser extent in phloem parenchyma (PP) contiguous with sieve elements (SE).

onstration of the transport of nitrogen fixed in mature leaf cavities toward the apical region, the transported compound(s) have not been identified and, as yet, there is no direct evidence that nitrogen is transferred from the host tissue to the generative *Anabaena* filaments. Kaplan and Peters (1981) also reported that whereas the nitrogen content and dry matter decreased with increasing leaf age, the carbon : nitrogen ratio increased. This is illustrated in Fig. 18, along with the relative nitrogenase activity of the excised segments from main stem axes.

These findings are consistent with the earlier suggestion that filaments of the endophyte that actively fix N_2, specifically those in mature cavities, might have diminished capability to metabolize the resulting ammonia

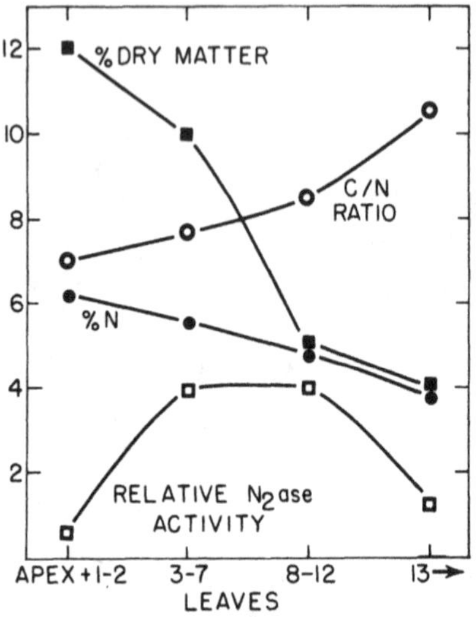

Fig. 18. The relative nitrogenase activity (□), percent dry matter (■), ratio of carbon to nitrogen (○), and the percent nitrogen (●) on a dry weight basis in segments of the *A. caroliniana* main stem axes bearing sequential groups of leaves. (Data plotted from Kaplan & Peters, 1981.)

(Ray et al., 1978). Moreover, this implies that the host exerts a control or regulation of metabolic processes in the endophyte as a function of the developmental gradient. In essence, the endophyte is caused to rapidly differentiate a disproportionate number of heterocysts and to exist in a state of metabolic idling, serving as an ammonium production facility in a manner reminiscent of the legume bacteroids.

The factors responsible for diminished cell division, greatly increased heterocyst differentiation, and diminished ability to assimilate the ammonia from N_2 fixation during the developmental profile in the *Azolla* endophyte are not yet resolved. There are, however, some insights. Although other plant-cyanobacterial symbioses do not exhibit a comparable developmental gradient, their endophytes do exhibit comparable heterocyst frequencies and the release of ammonia (Silvester, 1976; Rodgers & Stewart, 1977; Stewart, 1977). Furthermore, the endophytes generally have low or undetectable levels of GS (Stewart, 1977; Haselkorn, 1978). The low GS levels in the endophytes have been postulated as a biochemical mechanism explaining ammonia excretion (Stewart, 1977), and it has been suggested (Stewart, 1977; Haselkorn, 1978) that the host plants modify the endophytes' ammonia-assimilating pathways by producing effector substances that inhibit its GS activity or synthesis. This might also be related to increased heterocyst differentiation. Inhibitors of GS activity are known to moderately increase heterocyst frequencies in free-living cyanobacteria (Stewart & Rowell, 1975; Ladha et al., 1978). In the case of the *Azolla-Anabaena* association, Ray et al. (1978) suggested that the endophyte's GS activity might be associated primarily with the undifferentiated filaments in the apical portion of the stem. Subsequently, Haselkorn et al. (1980) employed an antibody against the purified GS from *Anabaena* 7120. They found that the antigen levels of the endophytic *Anabaena* were only 5%–10% of those observed in a free-living isolate and that the antigen concentration was greatest in the endophyte associated with younger leaves. Ray et al. (1978) found the GS levels in the endophytic *Anabaena* to be one-half those found in Newton's free-living isolate and *A. cylindrica*. However, they noted that epidermal hairs of the *Azolla* were present in the endophyte preparations and, although accounting for only 5% of the protein, a contribution by the hairs to GS activity was not excluded. In fact, it was stated that due to a possible contribution from the epidermal hairs the total GS activities attributed to the endophyte might be slightly high. Thus, there is reason to suspect a gradient in the endophyte's GS, decreasing in parallel with the differentiation of heterocysts as well as epidermal hairs in the leaf cavities. This situation would enable high heterocyst frequencies, low GS, and ammonia release by the endophyte with its assimilation by the host.

The other ammonia-assimilating enzyme, GDH, also merits consideration. Although not found in appreciable quantities in free-living N_2-fixing cyanobacteria or other symbiotic forms, appreciable levels of this enzyme were found in the *Azolla* endophyte, including a preparation from which the epidermal hairs were removed (Ray et al., 1978). GDH has an appreciably lower affinity for ammonia than does GS. Ray et al. (1978) postulated that the endophyte's GDH might be associated with those filaments occupying mature cavities and actively fixing N_2. This would result in the ammonia released by the endophyte normally being assimilated by the *Azolla* GS. However, the endophyte's GDH could have a regulatory role, enabling it to effectively reassimilate released ammonia at high intracavity ammonia concentrations. The actual amount of N_2 fixed by the endophyte that is utilized by the individual partners, and a complete understanding of nitrogen metabolism in the association as a function of the developmental profile, are subjects of current research.

The relative contribution of the individual partners to the association's total photosynthetic capability and the extent of interaction in fern-endophyte carbon metabolism are largely unknown. They may also vary as a function of the developmental profile, and it has been suggested that the endophyte might exhibit photoheterotrophic (Peters & Mayne, 1974b) or mixotrophic (Peters, 1975) metabolism. In most other plant-cyanobacterial symbioses, in which the endophyte simply exhibits a high heterocyst frequency with no developmental profile, the endophyte loses its capability to fix CO_2 and becomes dependent upon the plant for a source of fixed carbon (Stewart, 1978). A possible correlation between high heterocyst frequency and an exogenous carbon source is suggested by several observations. In lichens where the cyanobacterium is the only phycobiont, heterocyst frequencies are comparable to or slightly less than those occurring in the free-living cyanobacterium, that is, 4%. However, in a tripartite association between the same two organisms and a green alga, the cyanobacterium exhibits heterocyst frequencies in the range of 10%–30% (Millbank, 1977). This suggests that heterocyst frequencies in free-living cyanobacteria may be restricted (not regulated) by the availability of photosynthate from the vegetative cells and that to sustain functional heterocysts at frequencies of 15%–20% or more there may be an absolute requirement for an exogenous carbon source to maintain levels of reducing power. Thus, in the *Azolla-Anabaena* association we would postulate a transition from photoautotrophic metabolism in generative filaments to a photoheterotrophic or mixotrophic mode of metabolism with increasing differentiation of heterocysts. Ray et al. (1979) suggested that sucrose produced by the *Azolla* might serve as a reduced carbon source for the endophyte in mature leaf cavities with oxidative metabolism of sucrose conceivably providing reductant for N_2 fixation. The endophyte was found to possess glucose-6-phosphate dehydrogenase. In free-living cyanobacteria the activities of this enzyme are six to seven times higher in heterocysts than in vegetative cells (Apte et al., 1978). Glucose-6-phosphate can provide electrons to ferredoxin via glucose-6-phosphate dehydrogenase and ferredoxin $NADP^+$ oxidoreductase (Winkenbach & Wolk, 1973). Further, presumed isolates of the *Azolla* endophyte from *A. pinnata* (Becking, 1976) and *A. caroliniana* (Newton & Herman, 1979) are capable of dark heterotrophic growth and N_2 fixation on fructose. Finally, $^{14}CO_2$ pulse-chase and time-course studies showed that the *Azolla*, but not the endophytic *Anabaena*, synthesized sucrose (Ray et al., 1979). However, sucrose, glucose, and fructose have recently been identified as the major soluble di- and monosaccharides in the endophyte (Peters & Kaplan, 1981). This implies that sucrose found in the endophyte is synthesized by the *Azolla*. Current research involves the question of invertase activities in the endophyte as a function of development, an unambiguous demonstration that the sucrose in the endophyte is of host origin, and the possible involvement of simple hairs in the transport of fixed carbon into the leaf cavities.

Current hypothesis of functional organization

The salient aspects of our current interpretation of functional organization as it relates to *Azolla-Anabaena* interaction (see Kaplan & Peters, 1981; Calvert & Peters, 1981) are depicted in the schematic illustration

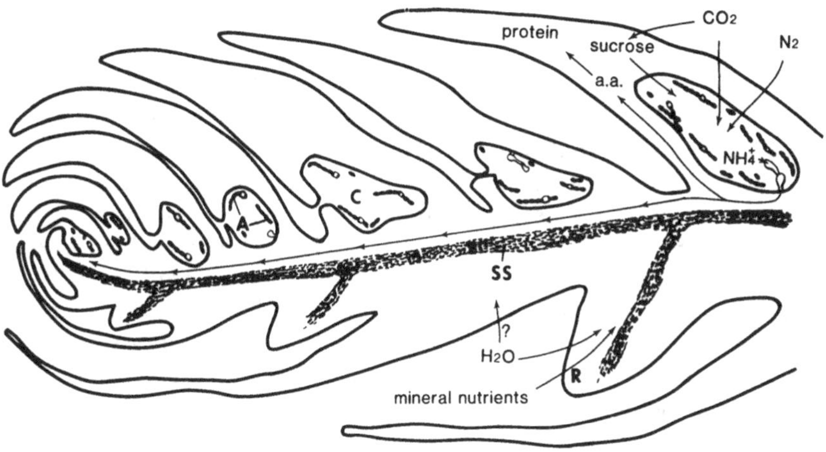

Fig. 19. A schematic illustration of a longitudinal section of the *Azolla-A. azollae* association depicting stages in the formation of the leaf cavity in the dorsal lobes, entrapment of the endophyte, and generalized physiological processes relevant to host-endophyte interaction. SS, stem stele; R, root. (The illustration is a modification of that in Peters, 1978.)

of a longitudinal section of the association (Fig. 19). The *Anabaena* colony associated with the shoot apex comprises generative *Anabaena* filaments. As with the filaments in subsequent stages of development, these filaments are effectively isolated from any source of nutrients other than those provided by the *Azolla*. However, they do not possess nitrogenase activity and, as Hill (1977) was the first to argue, the *Azolla* must provide them with a source of fixed nitrogen in addition to other nutritional requirements. The apical region has a higher nitrogen content than the rest of the plant, and it is continually supplied with nitrogen fixed by the *Anabaena* filaments in mature leaf cavities. The rapid differentiation of the primary branched hair and maturation of its terminal cells with transfer cell ultrastructure strongly suggest that this hair is the key to metabolite transfer, including nitrogenous compounds, from the *Azolla* to the apical *Anabaena* colony. Transfer cell ultrastructure has been implied to be involved with short-distance transport of solutes in other systems (Pate & Gunning, 1972).

As the generative *Anabaena* filaments are dissociated from the apical colony and partitioned into forming leaf cavities, several events occur simultaneously. The *Anabaena* filaments begin to enlarge, differentiate heterocysts, and exhibit nitrogenase activity. The terminal cells of the branched hairs senesce and a third cell type, the body cell, develops and differentiates transfer cell ultrastructure. Simple hairs begin to develop on the cavity walls and these also differentiate transfer cell ultrastructure. As leaf development proceeds, the heterocyst frequencies of the *Ana-*

baena increase dramatically and the filaments crowd around the periphery of the cavity in close proximity to the hairs.

It is our opinion that branched hairs may well be the principal sites of nitrogen interchange throughout the developmental sequence. Although it was suggested that the terminal cells would release nitrogen to the generative filaments, we suggest that the body cells, which develop as the terminal cells senesce, function in the uptake of fixed nitrogen released by the endophyte. These cells would logically contain high levels of ammonia-assimilating enzymes, specifically glutamine synthetase. The resulting amino acid(s) would move through the stalk cell of the branched hair and into the vascular system via the foliar trace. In our assessment the simple hairs would be less likely to play any direct role in the assimilation of ammonia released by the endophyte. Rather, they would provide a logical conduit for the transfer of photosynthate, presumably as sucrose, from the mesophyll into the mature leaf cavities. The photosynthate would then be metabolized by the endophyte. This would provide an additional source of reductant, satisfying the high demand for reductant imposed by high nitrogenase activity and high heterocyst frequency.

The completeness of the symbiotic state in the *Azolla-Anabaena* associations is exemplified by the retention of the endophyte throughout the *Azolla* life cycle. Among the N_2-fixing plant-prokaryote symbioses such continuity is unique to these associations. It is a remarkable feature in that it effectively precludes leaving the establishment of the symbiosis to chance. The extent of the mutualism is likewise remarkable. This is exemplified by the precise sequence of morphological and physiological events associated with the vegetative continuity of the symbiosis.

Within the association the *Azolla* provides the endophytic *Anabaena* with a highly specialized niche. Because the endophyte is always protected by the dorsal leaf lobe of the *Azolla*, either within a defined cavity or by virtue of overarching lobes in the apical region, it is never directly exposed to the natural aquatic and atmospheric environments. Mineral nutrients and water must be taken up through the fern prior to their reaching the cyanophyte. Light energy incident on the cyanophyte has been filtered by the fern leaf tissues. Although gases may freely diffuse into the leaf cavities, it is also possible that the gaseous composition of the leaf cavity is significantly different from the atmosphere. Thus, the endophytic *Anabaena* receives physical protection, required minerals and other nutrients, and adequate moisture – in general, a highly specialized environment created for it.

No other cyanobacteria are ever found in the *Azolla* leaf cavities. Although Mellor et al. (1981) have recently demonstrated the occurrence of an *Azolla* lectin, it remains to be established whether the lectin plays any role in the establishment and maintenance of the specificity of the symbiosis. In any event, the niche offered the endophyte not only frees it from competition with free-living cyanobacteria but may also provide a competitive advantage for its continued existence. It literally lifts the endophyte above the competition among other cyanobacteria and algae for solar energy.

The endophytic *Anabaena* necessarily pays a price for the niche it occupies. In essence it is used by the fern. Its growth rate, differentiation, ammonia-assimilating enzymes, and perhaps other properties, appear to be regulated by the fern such that the endophyte becomes an ammonia-producing facility. This use of the endophyte necessitates a cost to the *Azolla* in that it appears to provide the endophyte in mature cavities, that is, those actively fixing N_2, with photosynthate. It also transports nitrogen fixed in mature cavities to the apical tissues and, presumably, from the apical tissues to the generative *Anabaena* filaments external to them. The uptake, transport, and release of moisture, nutrients, fixed nitrogen, and fixed carbon undoubtedly result in some as yet unquantitated energy cost to the *Azolla*.

Research critical to future exploitation

The authors consider two areas of research critical to future exploitation of these associations. First, although there has been a number of reports of isolation and culture of the endophytic *Anabaena*, there has been no fully satisfactory demonstration of successful reestablishment of the symbiotic state with any isolate and an endophyte-free *Azolla*. Since it is difficult to acknowledge any of the isolates as the true endophyte until Koch's postulates have been successfully demonstrated, caution is required in comparative studies of the endophytic organism and such free-living isolates. Moreover, any meaningful genetic manipulation and/or cross-infection studies are effectively precluded until reliable techniques have been developed for inoculation and reinfection of endophyte-free plants. Second, much more detailed knowledge is required on the control of sporulation. The spores not only provide a potential means of preserving germ plasm, but they may also enable the creation of superior *Azolla* for use in agriculture through breeding programs.

References

Apte, S. K., Rowell, P., & Stewart, W. D. P. (1978). Electron donation to ferredoxin in heterocysts of the N_2 fixing alga, *Anabaena cylindrica*. *Proc. Royal Soc. Lond.* Series B. **200**, 1–25.

Ashton, P. J. (1974). The effect of some environmental factors on the growth of *Azolla filiculoides* Lam. *In: The Orange River Progress Report*, ed. E. M. V. Zinderen Bakker, Sr., pp. 123–38. Bloenfontein, South Africa: University of the Orange Free State.

Becking, J. H. (1976). Contributions of plant algal associations. *In: Proceedings of the 1st International Symposium on Nitrogen Fixation*, vol. 2, ed. W. E. Newton & C. J. Nyman. Pullman: Washington State University Press.

– (1979). Environmental requirements of *Azolla* for use in tropical rice production. *In: Nitrogen & Rice*, pp. 345–74. Los Banos, Laguana, Philippines: International Rice Research Institute.

Bierhorst, D. W. (1971). *The Morphology of Vascular Plants. Salviniales*. New York: Macmillan.

Bonnet, A. L.-M. (1957). Contribution à l'etude des hydropteridees. III. Recherches sur *Azolla filiculoides*. *Revue de Cytologie et de Biologie Vegetales* **18**, 1–86.

Bortels, H. (1940). Über die Bedeutung des Molybdäns für sticksoffbindende Nostocaceen. *Archiv für Mikrob.* **11**, 155–86.

Buckingham, K. W., Ela, W. S., Morris, J. G., & Goldman, C. R. (1978). Nutritive value of the nitrogenfixing aquatic fern, *Azolla filiculoides*. *J. of Agricultural & Food Chem.* **26**, 1230–4.

Calvert, H. E., & Peters, G. A. (1981). The *Azolla-Anabaena azollae* relationship. IX. Morphological analysis of leaf cavity hair populations. *New Phytol.* **89**, 327–35.

Campbell, D. H. (1893). On the development of *Azolla filiculoides* Lam. *Ann. of Bot. (London)* **7**, 155–87.

– (1895). *The Structure and Development of the Mosses and Ferns*. New York: Macmillan and Company.

Dao, T. T., & Tran, Q. T. (1979). Use of *Azolla* in rice production in Vietnam. *In: Nitrogen & Rice*, pp. 395–405. Los Banos, Laguna, Philippines: International Rice Research Institute.

Duncan, R. E. (1940). The cytology of sporangium development in *Azolla filiculoides*. *Bull. of the Torrey Bot. Club* **67**, 391–412.

Espinas, C. R., Berja, N. S., Del Rosario, D. C., & Watanabe, I. (1979). Environmental conditions affecting *Azolla* growth. *Greenfields* **9**, 14–19.

Florschütz, F. (1938). The two species of *Azolla* from the Pleistocene deposits of the Netherlands. *Recueil des Travaux Botaniques Néerlandais* **35**, 932–45.

Fogg, G. E., Stewart, W. D. P., Fay, P., & Walsby, A. E. (1973). *The Blue-Green Algae*. New York: Academic Press.

Gunning, B. E. S., Hughes, J. E., & Hardham, A. R. (1978). Formative and proliferative cell divisions, cell differentiation, and developmental changes in the meristem of *Azolla* roots. *Planta* **143**, 121–44.

Haselkorn, R. (1978). Heterocysts. *Ann. Rev. Plant Phys.* **29**, 319–44.

Haselkorn, R., Mazur, B., Orr, J., Rice, D., Wood, N., & Rippka, R. (1980). Heterocyst differentiation and nitrogen fixation in cyanobacteria (blue-green algae). *In: Nitrogen Fixation*, vol. 2, ed. W. E. Newton & W. H. Orme-Johnson, pp. 259–78. Baltimore: University Park Press.

Hill, D. J. (1975). The pattern of development of *Anabaena* in the *Azolla-Anabaena* symbiosis. *Planta* **133**, 237–42.

– (1977). The role of *Anabaena* in the *Azolla-Anabaena* symbiosis. *New Phytol.* **78**, 611–16.

Holst, R. W. (1977). Anthocyanins of *Azolla*. *Amer. Fern J.* **67**, 99–100.

Johnson, G. V., Mayeux, P. A., & Evans, H. J. (1966). A cobalt requirement for symbiotic growth of *Azolla filiculoides* in the absence of combined nitrogen. *Plant Phys.* **41**, 852–5.

Kaplan, D., & Peters, G. A. (1981). The *Azolla-Anabaena azollae* relationship. X. $^{15}N_2$ fixation and transport in main stem axes. *New Phytol.* **89**, 337–46.

Konar, R. N., & Kapoor, R. J. (1972). Anatomical studies on *Azolla pinnata*. *Phytomorphology* **22**, 211–23.
– (1974). Embryology of *Azolla pinnata*. *Phytomorphology* **24**, 228–61.
Ladha, J. K., Rowell, P., & Stewart, W. D. P. (1978). Effects of some amino acid analogues on growth and heterocyst formation in the bluegreen alga *Nostoc linckia*. *Proc. of the Indian Acad. of Science* (B) **81**, 127–33.
Liu, C. C. (1979). Use of *Azolla* in rice production in China. *In: Nitrogen & Rice*, pp. 375–94. Los Banos, Laguna, Philippines: International Rice Research Institute.
Lucas, R. C., & Duckett, J. G. (1980). A cytological study of the male and female sporocarps of the heterosporous fern *Azolla filiculoides* Lam. *New Phytol.* **85**, 409–18.
Lumpkin, T. A., & Plucknett, D. L. (1980). *Azolla*: botany, physiology, and use as a green manure. *Economic Botany* **34**, 111–53.
Mellor, R. B., Gadd, G. M., Rowell, P., & Stewart, W. D. P. (1981). A phytohaemagglutinin from the *Azolla-Anabaena* symbiosis. *Biochem. & Biophys. Research Commun.* **99**, 1348–53.
Millbank, J. W. (1974). Associations with bluegreen algae. *In: The Biology of Nitrogen Fixation*, ed. A. Quispel, pp. 238–64. New York: Elsevier.
– (1977). Lower plant associations. *In: A Treatise on Dinitrogen Fixation, Section III: Biology*, ed. R. W. F. Hardy & W. S. Silver, pp. 125–51. New York: Wiley.
Moore, A. W. (1969). *Azolla*: biology and agronomic significance. *Bot. Rev.* **35**, 17–34.
Newton, J. W., & Herman, A. I. (1979). Isolation of cyanobacteria from the aquatic fern, *Azolla*. *Arch. Microbiol.* **120**, 161–5.
Olsen, C. (1972). On biological nitrogen fixation in nature, particularly in bluegreen algae. *Comptes Rendus des Travaux du Laboratorie Carlsberg* **37**, 269–83.
Ott, F. D., & Petrik-Ott, A. J. (1973). *Azolla* and its occurrence in Virginia. *British Fern Gazette* **10**, 305–9.
Pate, J. S., & Gunning, B. E. S. (1972). Transfer cells. *Ann. Rev. of Plant Phys.* **23**, 173–96.
Peters, G. A. (1975). The *Azolla-Anabaena azollae* relationship. III. Studies on metabolic capabilities and a further characterization of the symbiont. *Arch. of Microbiol.* **103**, 113–22.
– (1976). Studies on the *Azolla-Anabaena azollae* symbiosis. *In: Proceedings of the 1st International Symposium on Nitrogen Fixation*, vol. 2, ed. W. E. Newton & C. J. Nyman, pp. 592–610. Pullman: Washington State University Press.
– (1977). The *Azolla-Anabaena azollae* symbiosis. *In: Genetic Engineering for Nitrogen Fixation*, ed. A. Hollaender, pp. 231–58. New York: Plenum.
– (1978). Blue-green algae and algal associations. *BioScience* **28**, 580–5.
Peters, G. A., & Kaplan, D. (1981). Soluble carbohydrate pool in the *Azolla-Anabaena* symbiosis. *Plant Phys.* **67**, S–37.
Peters, G. A., & Mayne, B. C. (1974a). The *Azolla, Anabaena azollae* relationship. I. Initial characterization of the association. *Plant Phys.* **53**, 813–19.
– (1974b). The *Azolla, Anabaena azollae* relationship II. Localization of nitrogenase activity as assayed by acetylene reduction. *Plant Phys.* **53**, 820–4.
Peters, G. A., Evans, W. R., & Toia, R. E., Jr. (1976). *Azolla-Anabaena azollae* relationship. IV. Photosynthetically driven, nitrogenase-catalyzed H_2 production. *Plant Phys.* **58**, 119–26.
Peters, G. A., Toia, R. E., Jr., & Lough, S. M. (1977). *Azolla-Anabaena azollae*

relationship. V. $^{15}N_2$ fixation, acetylene reduction, and H_2 production. *Plant Phys.* **59**, 1021–5.
Peters, G. A., Toia, R. E., Jr., Raveed, D., & Levine, N. J. (1978). The *Azolla-Anabaena azollae* relationship. VI. Morphological aspects of the association. *New Phytol.* **80**, 583–93.
Peters, G. A., Mayne, B. C., Ray, T. B., & Toia, R. E., Jr. (1979). Physiology and biochemistry of the *Azolla-Anabaena* symbiosis. *In: Nitrogen & Rice*, pp. 325–44. Los Banos, Laguna, Philippines: International Rice Research Institute.
Peters, G. A., Ray, T. B., Mayne, B. C., & Toia, R. E., Jr. (1980a). *Azolla-Anabaena* association: Morphological and physiological studies. *In: Nitrogen Fixation*, vol. 2, ed. W. E. Newton & W. H. Orme-Johnson, pp. 293–309. Baltimore: University Park Press.
Peters, G. A., Toia, R. E., Jr., Evans, W. R., Crist, D. K., Mayne, B. C., & Poole, R. E. (1980b). Characterization and comparisons of five N_2-fixing *Azolla-Anabaena* associations. I. Optimization of growth conditions for biomass increase and N content in a controlled environment. *Plant, Cell & Environment* **3**, 261–9.
Peters, G. A., Ito, O., Tyagi, V. V. S., & Kaplan, D. (1981a). Physiological studies on N_2 fixing *Azolla. In: Genetic Engineering of Symbiotic Nitrogen and Conservation of Fixed Nitrogen*, ed. J. M. Lyons, R. C. Valentine, D. A. Phillips, D. W. Rains, & R. C. Huffaker. New York: Plenum.
Peters, G. A., Ito, O., Tyagi, V. V. S., Mayne, B. C., Kaplan, D., & Calvert, H. E. (1981b). Photosynthesis and N_2 fixation in the *Azolla-Anabaena* symbiosis. *In: Current Perspectives in Nitrogen Fixation*, ed. A. H. Gibson & W. E. Newton, pp. 121–4. Canberra: Australian Academy of Science.
Pieterse, A. H., De Lange, L., & Van Vliet, J. P. (1977). A comparative study of *Azolla* in the Netherlands. *Acta Bot. Neerl.* **36**, 433–49.
Rao, H. S. (1936). The structure and life history of *Azolla pinnata*, R. Brown with remarks on the fossil history of the *Hydropterideae. Proc. of the Indian Acad. of Science* **2B**, 175–200.
Ray, T. B., Peters, G. A., Toia, R. E., Jr., & Mayne, B. C. (1978). *Azolla-Anabaena* relationship. VII. Distribution of ammonia-assimilating enzymes, protein, and chlorophyll between host and symbiont. *Plant Phys.* **62**, 463–7.
Ray, T. B., Mayne, B. C. Toia, R. E., Jr., & Peters, G. A. (1979). *Azolla-Anabaena* relationship. VIII. Photosynthetic characterization of the association and individual partners. *Plant Phys.* **64**, 791–5.
Rodgers, G. A., & Stewart, W. D. P. (1977). The cyanophytehepatic symbiosis. I. Morphology and physiology. *New Phytol.* **78**, 441–58.
Sculthorpe, C. D. (1967). *The Biology of Aquatic Vascular Plants*. London: Arnold.
Shi, T. C., Li, C. K., Wang, F. C., Chung, C. P., Chu, L. P., & Peters, G. A. (1981). Studies on nitrogen fixation and photosynthesis in *Azolla imbricata* (Roxb.) and *Azolla filiculoides* Lam. *Acta Botanica Sinica* **23**, 306–15 (in Chinese).
Singh, P. K. (1977). Multiplication and utilization of fern "*Azolla*" containing nitrogenfixing algal symbiont as green manure in rice culture. *Il Riso* **26**, 301–7.
– (1979). Symbiotic algal N_2 fixation and crop productivity. *In: Annual Reviews of Plant Sciences*, vol. 1, ed. C. P. Malik, pp. 37–65. New Delhi: Kalyain Publishers.
– (1980). Introduction of "Green *Azolla*" biofertilizer in India. *Current Science* **49**, 155–6.

Silvester, W. B. (1976). Endophyte adaptation in *Gunnera-Nostoc* symbiosis. *In: Symbiotic Nitrogen Fixation in Plants*, ed. P. S. Nutman, pp. 521–38. Cambridge: Cambridge University Press.

Smith, G. M. (1955). *Cryptogamic Botany*, vol. 2, 2nd ed. New York: McGraw-Hill.

Stewart, W. D. P. (1977). A botanical ramble among the bluegreen algae. *Brit. Phycol. J.* **12**, 89–115.

– (1978). Nitrogen fixing cyanobacteria and their associations with eukaryotic plants. *Endeavour* **2**, 170–9.

Stewart, W. D. P., & Rowell, P. (1975). Effects of L-methionine-DL-sulphoximine on the assimilation of newly fixed NH_3, acetylene reduction and heterocyst production in *Anabaena cylindrica. Biochem. & Biophys. Research Comm.* **65**, 846–56.

Subudhi, P. R., & Watanabe, I. (1979). Minimum level of phosphate in water for growth of *Azolla* determined by continuous flow culture. *Current Science* **48**, 1065–6.

Svenson, H. K. (1944). The new world species of *Azolla. Amer. Fern J.* **34**, 69–84.

Sweet, A., & Hills, L. V. (1971). A study of *Azolla pinnata* R. Brown. *Amer. Fern J.* **61**, 1–13.

Talley, S. N., & Rains, D. W. (1980a). *Azolla* as a nitrogen source for temperate rice. *In: Nitrogen Fixation*, vol. 2, ed. W. E. Newton & W. H. Orme-Johnson, pp. 311–20. Baltimore: University Park Press.

– (1980b). *Azolla filiculoides* Lam. as a fallow-season green manure for rice in temperate climate. *Agronomy J.* **72**, 11–18.

Talley, S. N., Talley, B. J., & Rains, D. W. (1977). Nitrogen fixation by *Azolla* in rice fields. *In: Genetic Engineering for Nitrogen Fixation*, ed. A. Hollaender, pp. 259–81. New York: Plenum.

Tyagi, V. V. S., Mayne, B. C., & Peters, G. A. (1980). Purification and initial characterization of phycobiliproteins from the endophytic cyanobacterium of *Azolla. Arch. of Microbiol.* **128**, 41–4.

Tyagi, V. V. S., Ray, T. B., Mayne, B. C., & Peters, G. A. (1981). The *Azolla-Anabaena azollae* relationship. XI. Phycobiliproteins in the action spectrum for nitrogenase-catalyzed acetylene reduction. *Plant Phys.* **68**, 1479–84.

Watanabe, I. (1978). *Azolla* and its use in lowland rice culture. *Soil & Microbe* (Japan), **20**, 110.

Watanabe, I., Espinas, C. R., Berja, N. S., & Alimagno, B. V. (1977). Utilization of the *Azolla-Anabaena* complex as a nitrogen fertilizer for rice. *International Rice Research Institute Research Paper Series* **11**, 115.

Watanabe, I., Berja, N. S., & Del Rosario, D. C. (1980). Growth of *Azolla* in paddy field as affected by phosphorus fertilizer. *Soil Science & Plant Nutrition* **26**, 301–7.

Whitton, B. A. (1973). Interactions with other organisms. *In: The Biology of the Blue-Green Algae*, ed. N. G. Carr & B. A. Whitton, pp. 415–33. Berkeley: University of California Press.

Winkenbach, F., & Wolk, C. P. (1973). Activities of enzymes of the oxidative and the reductive phosphate pathways in heterocysts of a blue-green alga. *Plant Phys.* **52**, 480–3.

Yatazawa, M., Tomomatsu, N., Hosoda, N., & Nunome, K. (1980). Nitrogen fixation in *Azolla-Anabaena* symbiosis as affected by mineral nutrient status. *Soil Science & Plant Nutrition* **26**, 415–26.

8

Algal-fungal relationships in lichens: recognition, synthesis, and development

VERNON AHMADJIAN
Department of Biology
Clark University
Worcester, MA 01610

JEROME B. JACOBS
St. Vincent Hospital and
University of Massachusetts Medical School
Worcester, MA 01610

Lichens are symbiotic associations of fungi and algae. The association results in a morphological transformation of the fungus, and sometimes the alga, and produces a unique growth form called the lichen thallus. About 20,000 species of lichens are distributed over a broad geographical range. They grow on and within many different substrates.

The ecological and evolutionary success of lichens has led to a popular image that they are premier examples of mutualism. This idea of a harmonious and cooperative relationship between algae and fungi is an attractive and reasonable assumption that is based on a presumably long period of coevolution between the symbionts. However, there is little experimental evidence to support the mutualistic hypothesis (Smith, 1980). The role of the phycobiont in the symbiosis is clear. It provides the mycobiont with sugar alcohols, which are used for nutrients. The role of the mycobiont is not clear. There is no evidence of transport of any substance from fungus to alga. Minerals needed by the alga can be absorbed directly from rainwater that passes over the thallus (Smith, 1980). The only benefit the fungus may provide to the alga is to shade it from excessive light, and even that possibility has not been proven.

A different picture of the lichen symbiosis may be equally valid. The fungus may be thought of as a biotrophic parasite and the relationship between the symbionts a balanced one. Parasitism of the alga may have slowed to a point where the percentage of cells killed is balanced by new cells added to the population by division of existing algal cells. Such a strategy has survival benefits for both symbionts and implies some type of regulatory process, that is, controlled parasitism. The hypothesis that lichen fungi are parasites on algae has supportive evidence. For example,

This research was supported by a grant to V. A. from the National Science Foundation (PCM-7921676). We thank Paul Stein and Lorraine A. Russell for technical assistance and Chicita F. Culberson for TLC and HPLC analyses.

haustoria, which are common in lichens, can be explained better in a parasitic than a mutualistic relationship. Haustoria do not seem to be involved in nutrient transfer (Collins & Farrar, 1978; Hessler & Peveling, 1978). They may be structures that have no role in the symbiosis, that is, vestigial remnants of pathogenic progenitors. Also, concentric bodies, which are either organelles of unknown function or a new type of mycovirus, occur in the hyphae of many lichens, and among unlichenized fungi they are found predominantly in plant pathogens. According to Chapman and Good (Chapter 9 of this volume), the alga *Cephaleuros* is parasitized by fungi that form foliicolous lichens.

Our theme in this chapter is that the lichen fungus is a parasite and the alga is the host. This reflects the beliefs of early investigators (Debary, 1866; Schwendener, 1868), some of whom considered the lichen relationship to be a controlled parasitism (Peirce, 1900; Elenkin, 1902; dePuymaly, 1967; Kershaw & Millbank, 1970).

Our ability to synthesize lichens axenically allows us to determine if one mycobiont can combine with different algae. Such tests have revealed the parasitic nature of some lichen fungi (Ahmadjian & Jacobs, 1981). How an algal host defends itself against a fungus is not clear but may be considered in the context of other host-parasite relationships where the host cells produce toxins that slow or stop the growth of the parasite (Albersheim & Valent, 1978). It seems paradoxical that a parasitic relationship can be so successful. A lichen is an example of two different organisms, alga and fungus, that have evolved a successful strategy for survival.

Lectins and lichens

Lectins (phytohemagglutinins) occur in different lichens (Barrett & Howe, 1968; Filho et al., 1971), and they are common also in other plants (Heslop-Harrison, 1978). The first evidence that they may be involved in recognition between lichen symbionts was reported by Lockhart et al. (1978), who isolated lectins from *Peltigera canina* and *Peltigera polydactyla*. These compounds agglutinated A, O, and B group erythrocytes and were shown to be of mycobiont origin. *P. polydactyla* lectin was tested for its ability to bind to 11 different blue-green algae, two of which were from symbiotic associations. The lectin bound strongly only to the *Nostoc* phycobiont of *P. canina* and to *Gloeocapsa alpicola*. The latter alga, contrary to what the authors stated, is not a lichen phycobiont (Ahmadjian, 1967). Curiously, the lectin did not bind well to *Nostoc sphaericum*, a symbiont isolated from the hornwort *Anthoceros punctatus* and a common phycobiont of lichens (Ahmadjian, 1967). Results of this study and similar results by Petit (1982) with *Peltigera horizontalis* were inconclusive in showing that lectins are involved in symbiont recognition in blue-green lichens.

Bubrick and Galun (1980) tested binding of the lectin concanavalin A to *Trebouxia* and *Pseudotrebouxia* phycobionts in culture and newly isolated from lichens. With one exception, the lectin bound to the cultured cells but not to those newly isolated from the thallus. The lectin did not bind to the *Trebouxia* phycobiont of *Cladonia convoluta* in either condition. The phycobionts of the six other lichens used in this study most likely belonged to *Pseudotrebouxia*. The authors observed unspecified changes in the cell walls of the phycobionts after they were isolated from the lichens. From our experience these changes include a thicker wall and an extracellular gelatinous matrix. Such a matrix, observed in scanning electron microscope studies of early lichen resynthesis, binds hyphae of the mycobiont.

We doubt that a lectin-mediated recognition system exists in lichens because such a system implies a close specificity between the symbionts. Synthesis studies have shown that hyphae are nonspecific: They bind to inert objects as well as to different algae. Moreover, the specificity between algal and fungal symbionts seems to be very broad, one species of phycobiont forming lichens with widely different mycobionts (Hildreth & Ahmadjian, 1981).

Recognition and resistance

Recent studies on lichen synthesis (Ahmadjian et al., 1980; Ahmadjian & Jacobs, 1981) again raised the possibility that some type of recognition occurs between algal and fungal symbionts. These studies revealed that the mycobiont *Cladonia cristatella* formed lichens with some algae, not others, and only partially lichenized some algae. Using mica strips on which to synthesize lichens, Ahmadjian and Jacobs (1981) found that the *C. cristatella* mycobiont formed squamules with five species of *Trebouxia* phycobionts other than its natural one and presquamules with another *Trebouxia* phycobiont and the free-living desert alga *Friedmannia israeliensis*. The fungus parasitized and eventually killed 17 different algal symbionts that belonged to the genera *Myrmecia, Nostoc, Pseudochlorella, Trebouxia*, and *Pseudotrebouxia*, as well as various nonlichenized algae.

Why does the mycobiont lichenize some species of *Trebouxia* and parasitize others? Taxonomic differences between the two groups are not evident. They both belong to the Chlorococcales and reproduce by zoosporogenesis. One possibility is that the parasitized species are more closely related to *Pseudotrebouxia* in having characteristics similar to those of vegetative cell division (desmoschisis) (Hildreth & Ahmadjian, 1981). These traits include a slow release of division products and a close contact of daughter cells with the parent cell wall. Another possibility may relate

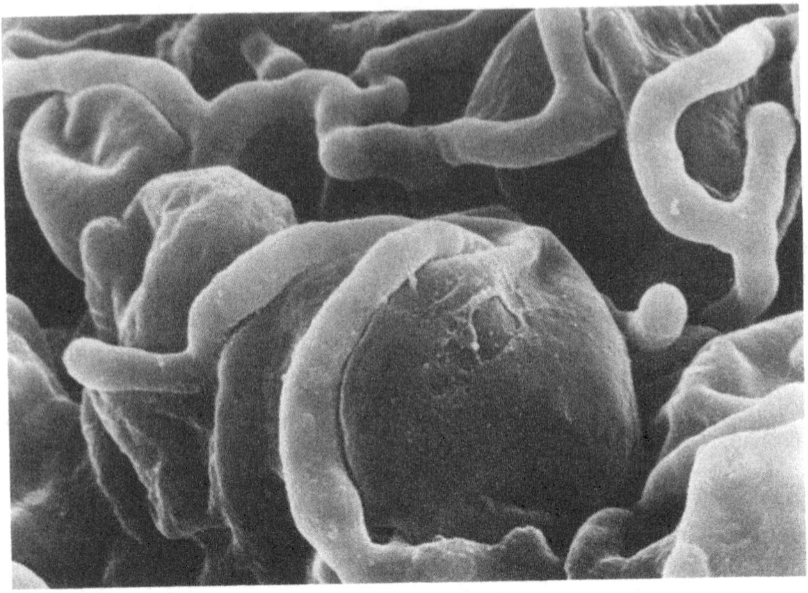

Fig. 1. Cell of *Trebouxia gelatinosa* phycobiont (*Parmelia caperata*) parasitized by hyphae of *C. cristatella* mycobiont in mica synthesis. 6,000 ×.

to the type of lichens from which the phycobionts were isolated. The compatible phycobionts came mostly from species of *Cladonia* or from lichens that belong to the Cladoniaceae, that is *Gymnoderma, Pilophoron*, and *Stereocaulon*. Other compatible phycobionts came from species of *Lecidea* and *Lepraria*, genera that may be distantly related to *Cladonia*. The mycobiont formed presquamules with *Trebouxia italiana*, an isolate from *Xanthoria parietina*, but it parasitized a *Pseudotrebouxia* isolate from another sample of *X. parietina*. The three species of *Trebouxia* that the fungus parasitized were phycobionts of *Parmelia* lichens (Fig. 1) and of *Xanthoria aureola*.

A surprising result was the lichenization of *F. israeliensis* (Fig. 2), an alga that was isolated from the Negev Desert (Chantanachat & Bold, 1962) and that has never been reported from a lichen association. *Friedmannia* and *Pseudotrebouxia* are in the same order, Chlorosarcinales, because they both show desmoschisis. *F. israeliensis* divides infrequently by desmoschisis and has parietal chloroplasts lacking pyrenoids (Deason et al., 1979). *Trebouxia erici*, the phycobiont of *C. cristatella*, has an axial chloroplast, fragments of which take a pronounced parietal position during zoosporogenesis (Hildreth & Ahmadjian, 1981). *Trebouxia* may have evolved from a free-living alga with parietal chloroplasts (Ahmadjian, 1970). *Pleurastrum terrestre*, also nonlichenized, is in the Chaetophorales. This alga is parasitized by the mycobiont even though its stages of zoo-

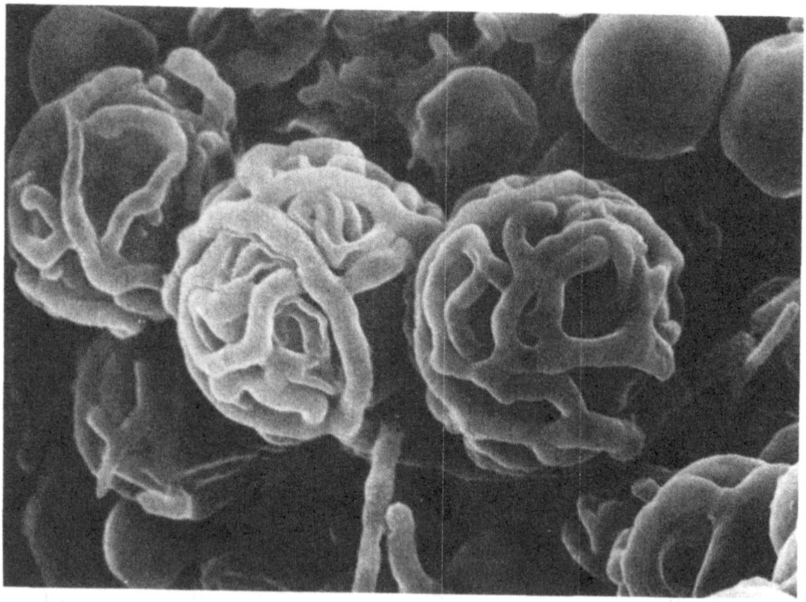

Fig. 2. Cells of *F. israeliensis* lichenized by hyphae of *C. cristatella* in mica synthesis. 2,300×.

sporogenesis and zoospores are similar to those of *Trebouxia* and *Friedmannia* (Molnar et al., 1975; Deason et al., 1979). *Trebouxia* has been reported to be a reduced form of *Pleurastrum* because of similarities between their pyrenoids (Molnar et al., 1975). The authors of this study, however, worked with *Trebouxia impressa*, an alga that has been transferred to *Pseudotrebouxia* (Archibald, 1975). Because the mycobiont did not lichenize the *Pseudotrebouxia* species used in our studies, it is not surprising that *Pleurastrum* was not lichized.

The association with *F. israeliensis* resulted only in presquamules. The fungus eventually killed the algal cells, which became filled with hyphae. This alga was more resistant to the mycobiont that *P. terrestre*. Hyphae of the mycobiont enveloped cells of *P. terrestre* but not as extensively as cells of phycobionts that they lichenized. The alga could not tolerate the fungus because the cells died and were filled with hyphae before presquamules formed. In the synthesis of *P. terrestre* and *F. israeliensis*, white patches of dead algal cells were present near green clumps of algae that were not yet killed by the fungus.

Why some algae are lichenized and others are not is unclear. If recognition exists between symbionts, it is a secondary phenomenon, occurring after the initial contacts. The situation in lichens may be similar to host-parasite relationships in higher plants (Albersheim & Anderson-Prouty, 1975; Albersheim & Valent, 1978), where host cells produce toxic compounds called phytoalexins, which stop or slow the growth of pathogens within the cell. These defensive compounds are stimulated by elicitors located on the cell wall of the pathogen. When an avirulent pathogen penetrates a host cell, receptor sites on the invaginated plasma membrane of the host activate elicitor molecules on the haustorial wall and phytoalexins are synthesized. If the elicitor is not recognized by the host cell, defensive compounds are not formed and the pathogen becomes virulent.

Haustoria are present in many lichens. How deeply the haustoria penetrate the algal cells has been correlated to the degree of thallus organization (Plessl, 1963; Galun et al., 1971). In general, lichens without a highly developed thallus have deeply penetrated algal cells, whereas lichens with well-developed thalli have algae with haustoria that do not extend far beyond the cell wall. In some genera of highly evolved lichens (e.g., *Peltigera* and *Sticta*) haustoria are not present.

The frequency of haustoria in lichens is a matter of dispute. Collins and Farrar (1978) reviewed published reports on lichen haustoria and found that the percentage of algal cells penetrated in different lichens ranged from zero to 60%. Recent studies have attempted to determine haustorial frequency by electron microscope observations. Such a method can be misleading. For example, Durrell (1967) reported that haustoria were absent in most of the 27 lichens he studied. In lichens where they occurred, only 1%–5% of the cells were infected. One lichen Durrell examined was *C. cristatella*. Our observations of this lichen (Ahmadjian & Jacobs, 1981) with a light microscope showed that about 60% of the algal population was penetrated by haustoria.

If we assume that the evolution of lichens has led to a more complex thallus organization, then we might hypothesize that advanced lichens have algal cells with smaller haustoria because of a greater recognition by the host (phycobiont) of antigens present on the walls of the pathogen (mycobiont). Such recognition leads to a better defensive strategy by the algal host, namely, production of phytoalexins, which limit the growth of the pathogen.

Symbiotic modifications of phycobionts

Algal symbionts of lichens change when they are removed from the symbiosis. They release smaller amounts of carbohydrates and show cell wall modifications. Green and Smith (1974) found that isolated *Trebouxia* cells rapidly lost their ability to excrete ribitol and incorporated more labeled carbon into ethanol-insoluble materials. Drew and Smith (1967) reported similar behavior for a *Nostoc* phycobiont that excreted glucose. Some cultured phycobionts develop a prominent extracellular sheath. The reason why isolated phycobionts excrete less carbohydrates is not

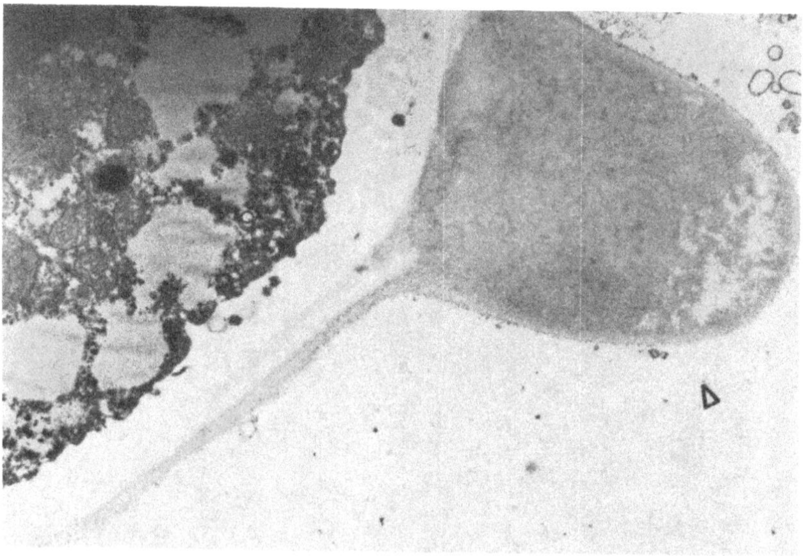

Fig. 3. Cultured cell of *A. calcarea* phycobiont with outgrowth of wall material (arrow). 13,000×.

known. Smith (1967) believes that cultured phycobionts retain carbohydrates for cell wall synthesis, which might explain why they divide more rapidly than phycobionts in the lichen thallus.

In a lichen the alga supplies polyols to the mycobiont, which uses them as a source of carbon and as a means of maintaining a high carbohydrate pool. Farrar (1976a) feels that this pool buffers a lichen against respiratory losses caused by alternate drying and wetting cyles. He estimated that 70%–80% of the carbon fixed by a phycobiont is transferred to the mycobiont (Farrar, 1976b).

Differences in physiology, ultrastructure, and antigenic properties between thallus and cultured phycobionts have been shown by Richardson (1973), Peveling (1973), and Bubrick et al. (1982), respectively. Our ultrastructural observations support the contention by Smith (1967) that cultured algae use carbohydrates, normally excreted by lichenized algae, for synthesis of cell wall material.

Cells of the cultured phycobiont of *Aspicilia calcarea* (= *Pseudotrebouxia* sp.) showed irregular outgrowths of wall material (Fig. 3). These outgrowths appeared to be part of an outer layer of the cell wall. The inner layer of the wall was thicker and less electron dense. In the lichen thallus the algal cells had an outer thin, irregular electron-dense layer, but this seemed to be part of a general, intercellular matrix between the symbionts. A two-layered cell wall was seen also in the cultured phycobiont of *Omphalodium arizonicum* (= *Pseudotrebouxia* sp.) but out-

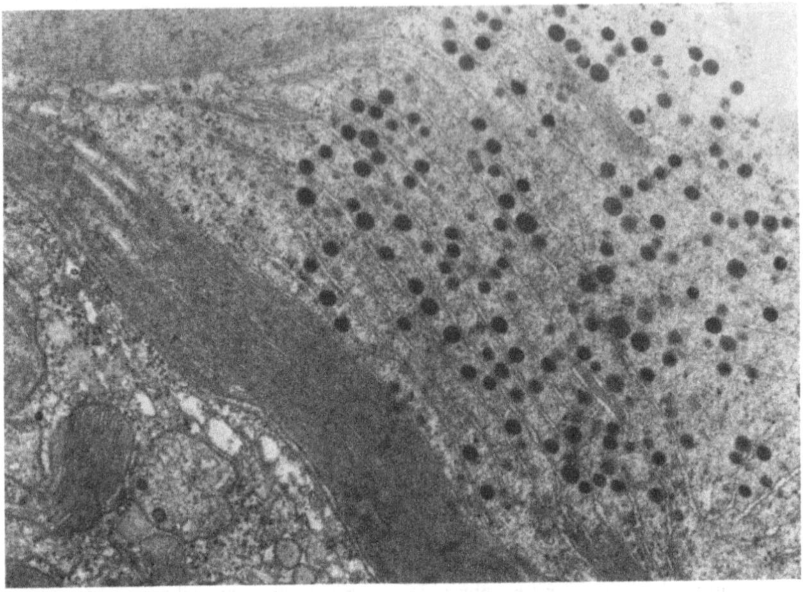

Fig. 4. Parallel thylakoids, lined with pyrenoglobuli, in pyrenoid of cultured cell of *O. arizonicum* phycobiont. 27,000×.

growths were not observed. A diffuse, granular sheath was seen around many cells. In this phycobiont, the wall of the cultured cells was larger (0.4–1.7 μm) than the wall of the thallus algae (0.3–0.5 μm). Outgrowths of wall material in the cultured *A. calcarea* alga were associated with dictyosomes and rough endoplasmic reticulum, structures that were common also in the cultured cells of *O. arizonicum* but rarely observed in either alga in the lichenized condition.

The chloroplast of both lichenized algae filled most of the cell volume, the cytoplasm being restricted to a thin rim around the cell periphery. In contrast, the chloroplast of the cultured cells was smaller and a wider band of cytoplasm was visible. In the lichenized phycobiont of *O. arizonicum*, the thylakoids were densest in the marginal lobes of the chloroplast. The cultured cells of this phycobiont had thylakoids that were uniformly distributed throughout the chloroplast. Starch was much more abundant in the cultured algal cells of both lichens. The starch granules were often large and elongated and were not associated in any particular pattern with the pyrenoid; rather, they were found scattered irregularly between the chloroplast thylakoids.

The pyrenoid matrix was granular in both the lichenized and cultured algae, but the pattern of thylakoid arrangement in the pyrenoids differed. Parallel rows of thylakoids in some pyrenoids of the cultured phycobiont of *O. arizonicum* (Fig. 4) and the vesiculated pattern in other cells were

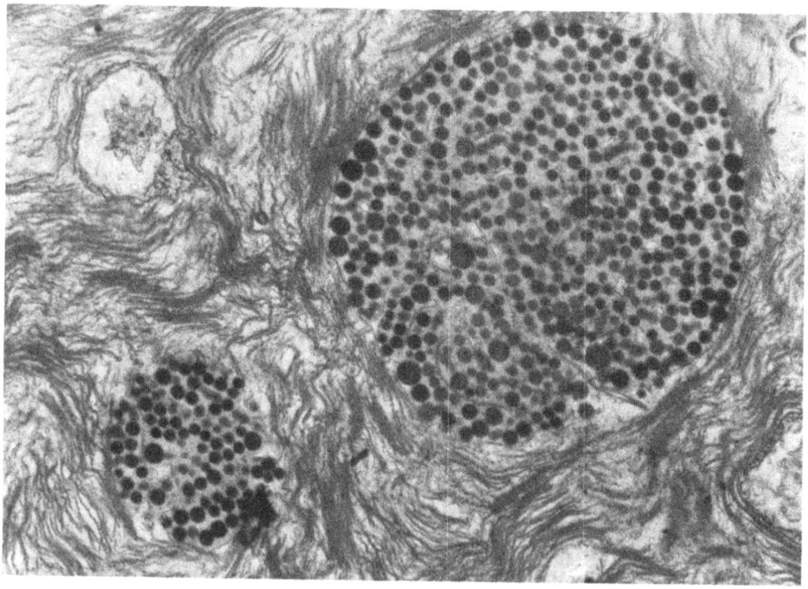

Fig. 5. Curved and dilated thylakoids in pyrenoid of lichenized cell of *O. arizonicum* phycobiont. 13,000×.

not seen in lichenized cells of this phycobiont, which had few randomly curved and often dilated thylakoids (Fig. 5). In *A. calcarea*, thylakoids of many of the cultured cells appeared dilated as they coursed through the pyrenoid matrix.

The lichenized algae had more pyrenoglobuli. They were up to 0.3 μm in diameter in *O. arizonicum* whereas the largest pyrenoglobuli in the cultured cells were about 0.1 μm, with most of them about half that size. In the lichenized cells of *A. calcarea*, most of the pyrenoglobuli aligned both sides of the thylakoid membranes in the pyrenoid. In *O. arizonicum*, there seemed to be many free-lying pyrenoglobuli, but it is not known how many of them were aligned with subtending membranes (Fisher & Lang, 1971).

Cells of *A. calcarea* had only one pyrenoid. The thallus algae in *O. arizonicum* had multiple pyrenoids that appeared to be of different sizes and shapes. The pyrenoid matrix was generally darker than the chloroplast stroma. Cultured cells of *O. arizonicum* generally had only one pyrenoid. The pyrenoid in some chloroplasts developed a vesiculation that completely obscured or destroyed the pyrenoid matrix (Fig. 6) and enlarged to the outer limits of the chloroplast. Pyrenoglobuli were not present in these vesiculated pyrenoids.

Numerous mitochondria were present in the lichenized and cultured phycobionts of both lichens. The mitochondria were usually oval but

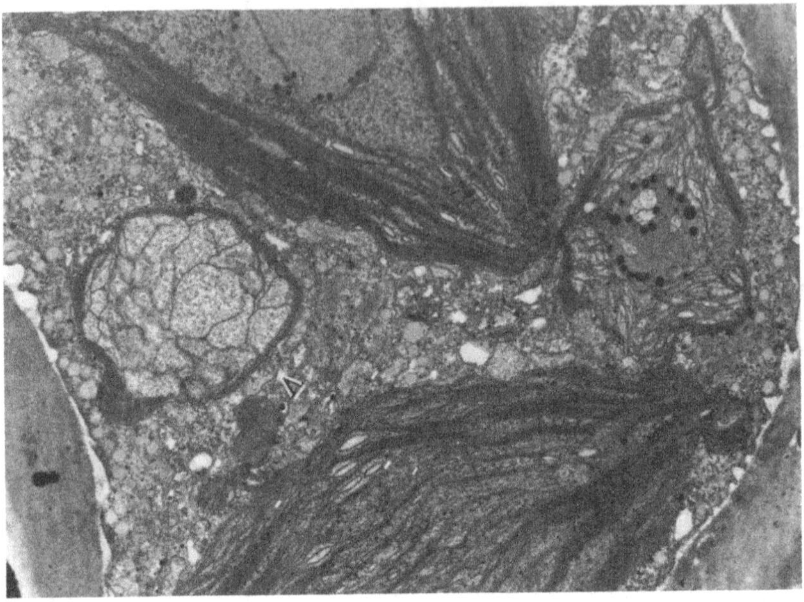

Fig. 6. Vesiculated pyrenoid (arrow) in cultured cell of *O. arizonicum* phycobiont. 13,000×.

sometimes irregular. They tended to aggregate along the outer margin of the chloroplast of the cultured algae. Division stages were more common in the cultured algae.

Storage droplets, presumably lipid, were larger and more numerous in the cultured algae, and the plasma membrane also was more irregular and in some instances deeply invaginated.

A contention by "whole-lichen" physiologists has been that the behavior of algal symbionts in culture is abnormal. Although such behavior may differ from that in symbiosis, it is not necessarily abnormal. Rather, it may reflect a presymbiotic condition that is necessary to attain before lichenization. One aspect of this presymbiotic state may be production of an extracellular material that binds the symbionts together. Ahmadjian et al. (1978) reported such a matrix around cells of the phycobiont (*Trebouxia glomerata*) of *Huilia albocaerulescens*, and we saw a similar matrix around cells of *T. erici*, the phycobiont of *C. cristatella*, both in the natural lichen (Fig. 7) and in the synthetic thalli. Recent studies have shown that de novo lichenization probably takes place in nature and that free-living phycobiont populations could originate from zoospores that escape from a lichen thallus (Ahmadjian et al., 1980; Culberson & Ahmadjian, 1980; Slocum et al., 1980). Thus, an algal symbiont could move from a lichenized to a free-living state and back to a lichenized state, and its phys-

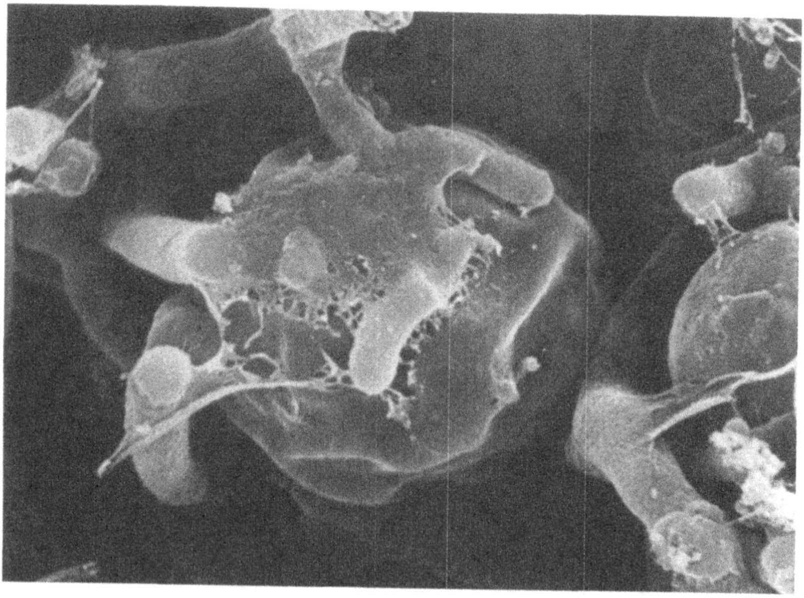

Fig. 7. Matrix around hyphae and phycobiont in natural thallus of *C. cristatella*. 7,000×.

iological state could shift accordingly. In the lichenized condition, a phycobiont's metabolism and physiology are directed toward the nutritional needs of the mycobiont and supplying its carbohydrate pool; the mycobiont constitutes the bulk of the thallus. However, in a nonsymbiotic condition the behavior of the phycobiont changes so that instead of releasing most of its photosynthetic products the symbiont retains these compounds for internal uses, one of which might be to synthesize the extracellular matrix that will develop its receptivity for a lichen fungus. The common occurrence of dictyosomes and rough endoplasmic reticulum in cultured cells of the phycobiont indicates that they secrete more substances than lichenized algae.

The greater abundance of pyrenoglobuli in lichenized cells compared to cells cultured in organic media has been noted by earlier investigators. Jacobs and Ahmadjian (1971) found that pyrenoglobuli in lichenized cells and cells grown in inorganic medium were the same size but those of cells grown in organic medium were half the normal size. Pyrenoglobuli are lipid-containing structures that presumably function as storage depots. They are similar to plastoglobuli, which, according to Lichtenthaler (1968), store excess lamellar lipids. Lichtenthaler noted an inverse relationship between the number and volume of plastoglobuli and thylakoid synthesis. Fisher and Lang (1971) related pyrenoglobuli to membrane metabolism in *T. erici* and found that at low light levels the cells

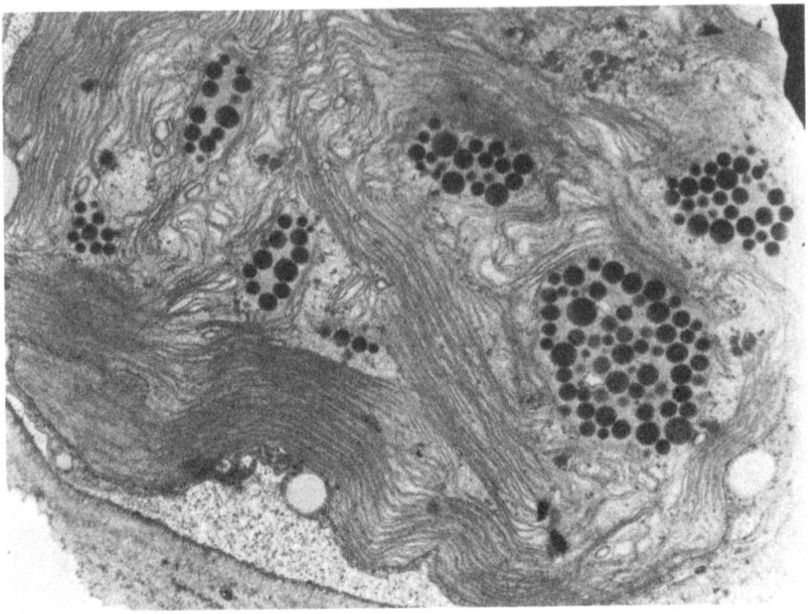

Fig. 8. Multiple pyrenoids in lichenized cell of *O. arizonicum* phycobiont. 13,000×.

had more pyrenoglobuli and fewer membranes than cells grown at high light intensities. The growth rate was more rapid at the highest light level used in this study.

The larger size and greater abundance of pyrenoglobuli in lichenized algae may be due to the slow division of these cells. Slocum et al. (1980) observed only a few division stages among the phycobiont population of *Parmelia caperta*. In September they reported only one dividing cell out of 184 randomly selected cells. In May, only 13% of the cells in one sample were dividing. Since much of the volume of lichenized *Trebouxia* or *Pseudotrebouxia* cells consists of the chloroplast, their slow division rate would favor an increase in pyrenoglobuli because lipids would not be needed for thylakoid synthesis as much as in the cultured algae. Pyrenoglobuli have been found also in nonlichenized marine algae such as *Monostroma groenlandicum* and *Capsosiphon fulvescens* (Hori, 1973) and in *Chlorella variegata*, a facultatively heterotrophic species (Silverberg & Sawa, 1974).

The differences in pyrenoid structure and number between lichenized and cultured algae relate to the physiological state of the algae. The phycobiont of *O. arizonicum* has many pyrenoids in the lichenized state (Fig. 8) whereas the cultured cells usually have only one pyrenoid. This discrepancy may be due to differences in the division rates. In the lichen alga very few division stages were seen, which indicates that pyrenoid multiplication is not coordinated with division of the chloroplast.

Synthesis and development of *C. cristatella*

Axenic synthesis

History. The first true axenic syntheses of lichens were made by E. A. Thomas (1939). He inoculated fungal and algal symbionts onto segments of elder pith that were embedded at one end in mineral agar contained in 150-cc Erlenmeyer flasks. Soredia were common in his cultures and in two flasks of the symbionts of *C. pyxidata* f. *chlorophaea* he observed squamules and podetia. Surprisingly, he could not duplicate this latter result in any of the additional 800 flasks he inoculated with the same symbionts.

The axenic syntheses reported by Bonnier (1889) have been criticized by later workers (Thomas, 1939; Quispel, 1943–5; Ahmadjian, 1962). His technique of inoculating the symbionts onto sterilized pieces of bark or stone suspended over water in a closed flask was not appreciably different from the successful methods reported by Thomas and ourselves. However, his use of algae obtained directly from tree bark can be criticized because these algae most likely were not phycobionts and they were not isolated into axenic cultures. Bonnier also conducted syntheses in chambers through which a current of air was passed. He reported that lichen thalli developed in several months to several years and that fruits appeared after 2 years.

Bartusch (1931) and Lange-de la Camp (1933) attempted axenic syntheses of *X. parietina* on fragments of clay and beerwort agar. The mixed growth of the symbionts did not develop beyond the initial contacts.

Artificial synthesis of *C. cristatella* was achieved by Ahmadjian (1966) on fragments of bark and pith that were partially embedded in a nutrient agar. The substrates were sown with spores of the mycobiont and cells of the phycobiont. Pycnidia and immature apothecia were produced by the resulting fungal colonies. Some older fungal colonies placed onto the wood fragments and coated with algal cells formed presquamules and a few squamules as they dried. This method of synthesis, like that of Thomas, gave inconsistent results, especially with regard to the formation of immature fruits. Most likely this was due to the fact that the fungal inoculum consisted of many spores and the resulting colonies were a mixture of genetic strains.

Synthesis flasks. Recently, we achieved axenic syntheses of *C. cristatella* (Ahmadjian & Jacobs, 1981) by means of a technique similar to that used by Thomas (1939). Syntheses were carried out in 125–ml Erlenmeyer flasks with 20 ml of a 2% and nonabsorbent cotton plugs. A newly

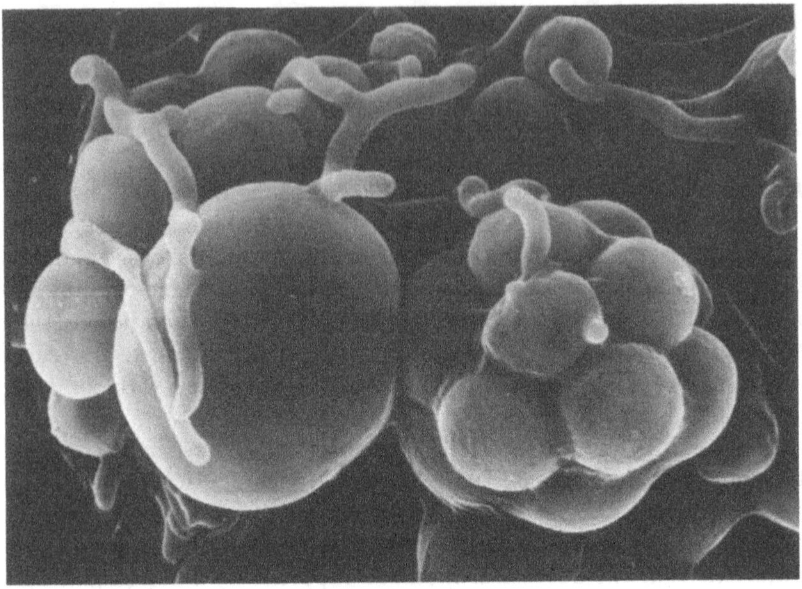

Fig. 9. *C. cristatella* mycobiont × *T. erici* phycobiont (*C. cristatella*) synthesis. Note early attachment of hypha to algal cell. 4,000×.

cleaved mica strip (1 × 7.5 cm) that had been soaked in a solution of Bold's Basal Medium X3N and dried was positioned so that one end of the strip was embedded in the agar and the other end rested against the side of the flask at about 45° angle. The symbionts were inoculated onto the cleaved surface of the mica. The hydrophilic nature of this surface probably bound the fungal mass. After inoculation, the cotton plugs were covered with aluminum foil secured tightly around the neck of the flask by means of a rubber band. Relative humidity in the closed flask was about 95%.

Flasks in which the mycobiont had grown were decanted of media. The mycelium was then washed by adding distilled water to the flasks. The water was decanted and the fungus was transferred to a petri dish. The mycelium was lightly dried with filter pads and mixed with algae removed from agar slants with a sterile spatula. The mixed symbionts were then gently pressed onto the mica surface with the spatula.

Early interactions between symbionts. A gelatinous matrix around the algal cells bound the fungal hyphae. Some algal cells had only a thin sheath (Fig. 9) whereas other cells were thickly coated. During the later stages of lichenization, the mycobiont also produced a gelatinous matrix that enveloped the hyphae and formed a protective coating for the squamules.

The fungus was not specific in its early attachment to algae. It attached

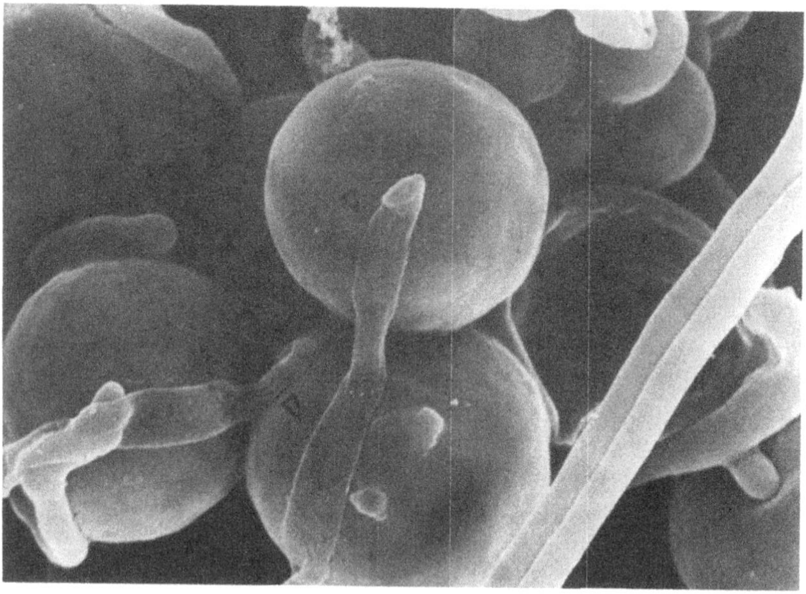

Fig. 10. *R. chrysoleuca* mycobiont × *Trebouxia excentrica* phycobiont (*Huilia tuberculosa*) synthesis. Note flattened appressoria (arrows). 4,000 ×.

to different algae and even to glass beads (10–15-μm diameter) that were substituted for algal cells in some synthesis cultures. Hyphae followed the contour of the beads but did not envelop them beyond the initial contacts.

Well-defined appressoria were not formed by *C. cristatella* mycobiont. In contrast, the mycobionts *Huilia albocaerulescens* and *Rhizoplaca chrysoleuca* formed pronounced appressoria in our synthesis cultures (Fig. 10). These structures resembled collapsed hyphal cells and their function was not clear. The appressoria did not terminate a hyphal filament; rather, the hypha grew out from the tip of an appressorium, resumed its cylindrical shape, and then formed another appressorium as it encircled a neighboring algal cell. These successive appressoria may be how the fungus secures algae to form the autotrophic population necessary for a lichen.

Haustoria were common in all our syntheses. Our SEM observations revealed that the haustoria pushed into the algal cell (Fig. 11). This observation conflicts with our earlier suggestion that the haustoria of *C. cristatella* penetrate the algal cells by enzymatic digestion (Jacobs & Ahmadjian, 1971). It is possible that enzymes may soften the wall to a point where it yields to haustorial pressure. Haustoria were produced near the tips of hyphae as well as farther back. The percentage of algal cells penetrated by haustoria in resynthesized squamules (4 months old) was about

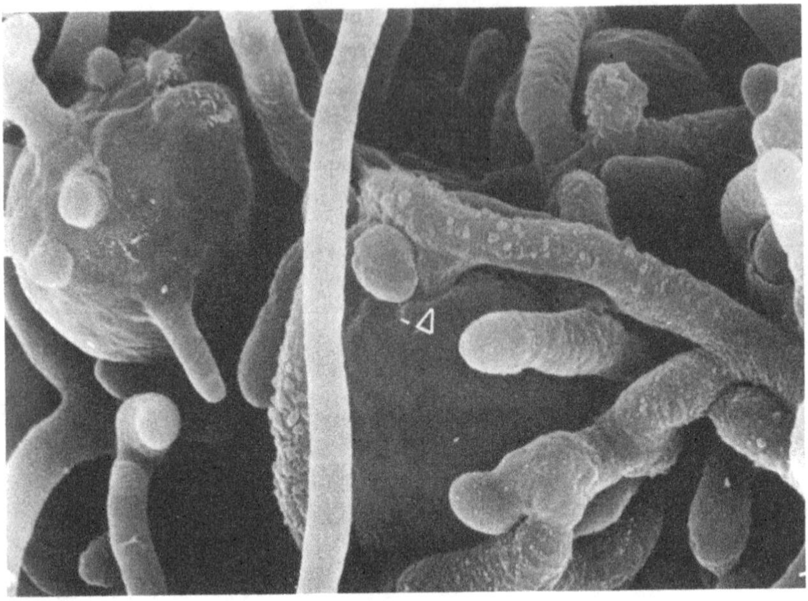

Fig. 11. *C. cristatella* mycobiont × *Trebouxia excentrica* phycobiont (*Cladonia subtenuis*) synthesis. Note haustorium (arrow) and crystals on hyphae and algal cell. 6,000×.

half that of the natural squamules (i.e., 60% vs. 26%). Despite this high frequency of infection, the algal population in both types of squamules appeared healthy.

Development of squamules

Our synthesis cultures produced many small (0.5–1.0 mm long) squamules that generally were crowded together (Fig. 12). They formed on the surface of the mixed symbionts and directly on the mica from aerial hyphae and algal cells. Mature squamules were upraised at one end.

After the fungal mass was placed onto the mica strips, it dried in several days to a doughy consistency. The dried fungus had a pockmarked surface from which aerial hyphae grew out and attached to algal cells (Fig. 13). These receptive hyphae must differ from those produced in liquid culture. Lichenization does not occur if the fungus is kept wet. An ultrastructural study comparing cell walls of aerial hyphae with those of hyphae cultured in liquid medium would be interesting. Differences between these hyphae might give clues to the specific mechanisms involved in early symbiotic interactions.

Initial contacts of algae by fungal hyphae were followed by a proliferation of hyphal branches around the algal cells (Fig. 14). The branching

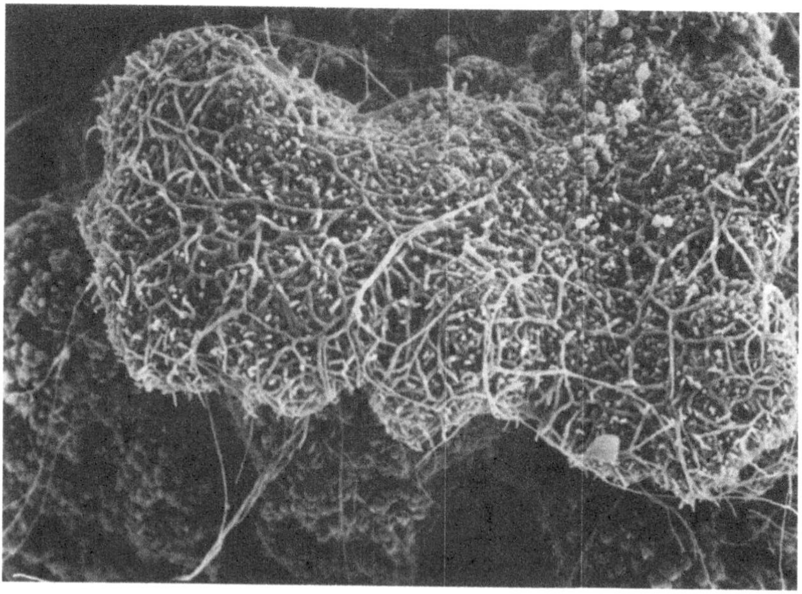

Fig. 12. *C. cristatella* mycobiont × *T. erici* phycobiont synthesis. Developing squamules. 300×.

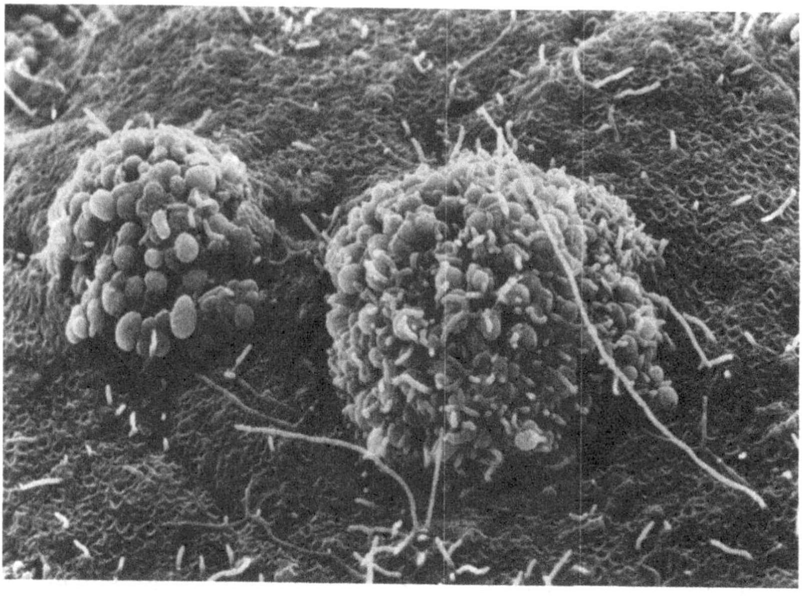

Fig. 13. Aerial hyphae of *C. cristatella* mycobiont growing out from pockmarked surface of original fungal inoculum. These receptive hyphae are shown enveloping algal cells. 800×.

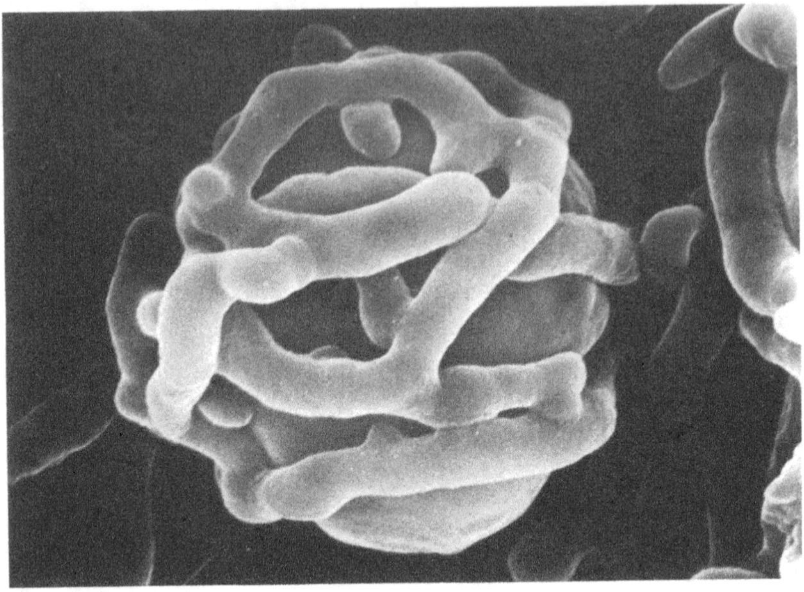

Fig. 14. Hyphal branches of *C. cristatella* mycobiont around an algal cell of *Trebouxia erici.* 7,000×.

did not follow a specific pattern. Many algal cells were covered almost completely by hyphae. In contrast, we found that algal cells in the natural lichen had a much greater exposed surface area.

The next stage of development was the formation of presquamules, sorediallike groups of algal cells enveloped by hyphae (Fig. 15). Such groups differed in size and were scattered throughout the culture. We could not determine if the algae divided after they were in a presquamule or whether the larger presquamules reflected the ability of hyphae to accumulate more algae. Greenhalgh and Anglesea (1979) observed that the population of the phycobiont in *Parmelia saxatilis* increased because of hyphae that penetrated the algal sporangia and enveloped the aplanospores.

Sometimes, after inoculation of the mixed symbionts, if the fungal mass was still wet, it slid down the mica and settled partially on the agar. Because of the excessive moisture in this location, the fungus did not dry and only presquamules formed on the upper surface of the fungal mass.

A superficial network of hyphae grew over the soredia and interlocked them into a common mass. As the network became thicker, gelatinization of walls formed an extracellular matrix that bound the hyphae together (Fig. 16). Thickened and fused hyphae occurred first in small patches and later over a broad area of the mixed culture. Directly below the

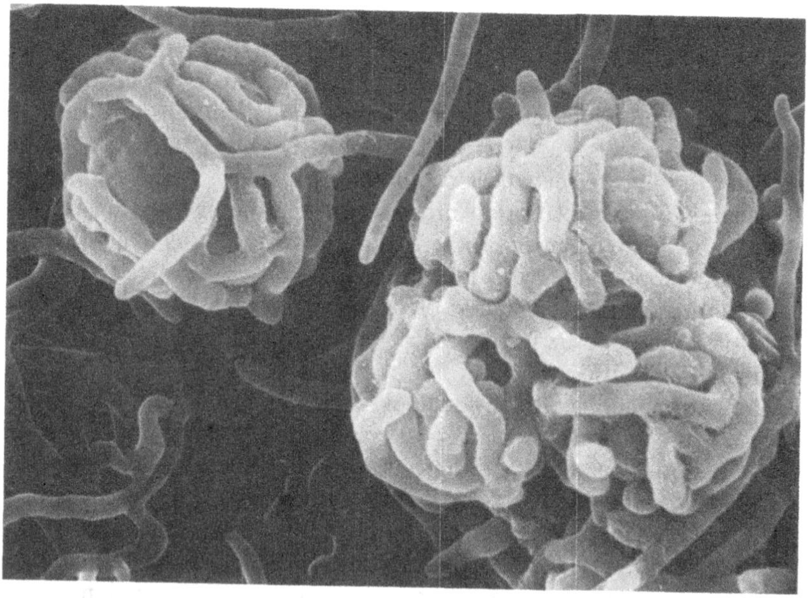

Fig. 15. Presquamules, sorediallike groups of algal cells, and hyphae of synthetic *C. cristatella*. 3,000×.

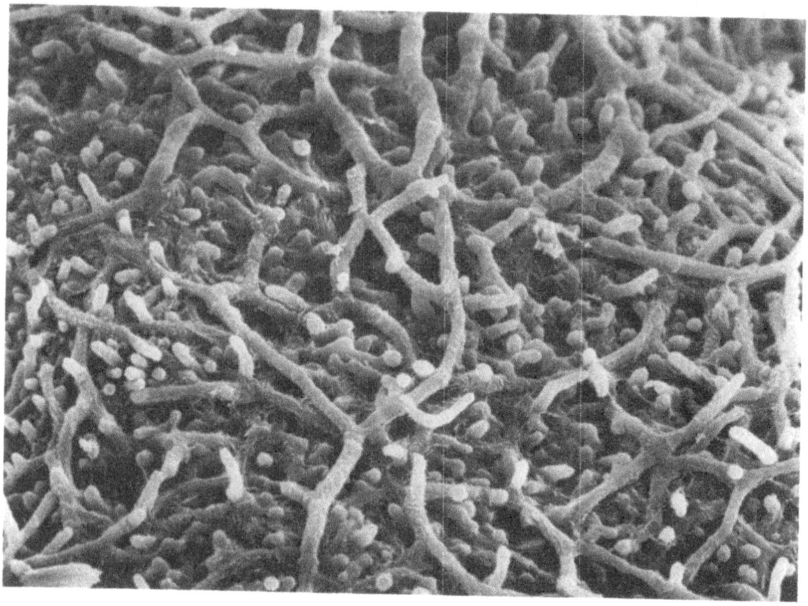

Fig. 16. *C. cristatella* mycobiont × *Trebouxia excentrica* phycobiont (*C. subtenuis*) synthesis. Network of hyphae on surface of squamule showing extra-hyphal matrix. 9,000×.

upper hyphal network, other hyphae became heavily gelatinized and compressed and formed the upper cortex of the squamule. Further development resulted in the formation of the algal layer and medulla.

In some of our synthesis cultures, hyphae and algal cells had irregular crystalline deposits on their outer surfaces (Fig. 11). It is not known whether these were secondary compounds produced by the synthesized squamules. Since the fixation procedure for scanning electron microscopy involved dehydration of the culture in ethanol, it is likely that any chemical compounds in the squamules would have dissolved in the alcohol and perhaps precipitated out at a later stage of the preparation. Thus, if these deposits were lichen crystals, their locations on the symbionts that we observed might not be natural ones.

The natural and synthetic squamules were identical. The only difference was the presence of numerous rod-shaped bacteria on the surface of the natural squamules and their absence on the synthetic squamules.

On mica strips that had been coated with malt-yeast extract medium and dried, the mixed symbionts produced mostly presquamules after 3 months of culture. A few squamules formed but the growth of the fungus on the nutrient medium was too strong to support lichenization, a phenomenon noted many times by earlier investigators. Washing the fungal mycelium with distilled water before mixing with algae removed culture medium (malt-yeast extract) and facilitated the formation of squamules.

Fungi fragmented with algae in a blender, washed and centrifuged (8,000 rpm) several times, and then inoculated onto mica strips and incubated formed only presquamules after 2 months. There was no mycelial growth of the fungus 3 weeks after inoculation. The hyphae may have been fragmented too finely and perhaps could not draw on reserve stores of food from older hyphal parts to initiate new growth. Aerial hyphae were produced more rapidly from undisturbed fungal mats.

Concentric bodies have not been observed in the hyphae of cultured isolates of lichenized and nonlichenized fungi that normally contain these structures. The mycobiont *C. cristatella* was cultured from ascospores, which apprently do not contain concentric bodies although they have been found in the spores of other lichens (Ascaso & Galvan, 1975). These bodies reappeared in squamules of *C. cristatella* from soil syntheses but not in the frequency found in the natural lichen (Ahmadjian et al., 1980). We thought they might be involved in the development and chemistry of a lichen thallus but this hypothesis is incorrect. We did not find concentric bodies in axenic squamules of *C. cristatella*. Such squamules were identical in morphology and chemistry to those formed in soil syntheses and to the natural squamules. The absence of these bodies in an axenic lichen supports the idea that concentric bodies represent some stage of a virus, and their presence in lichen fungi depends on an infectious source such as soil (M. Zavada, personal communication).

Chemistry of resynthesized lichens

Most lichen products appear to be the result of symbiotic interactions. According to Culberson and Ahmadjian (1980), synthesis of many lichen products may be due to algal inhibition of fungal enzymes. Mycobionts do not produce in culture the same chemical compounds that they do in symbiosis, with some exceptions. For example, analyses of 50 *C. cristatella* mycobiont strains (single-spore isolates) cultured in Lilly and Barnett medium showed that 10% (five) produced didymic acid, a dibenzofuran, found in many, but not all, natural populations of the lichen. This compound was found also in some of the synthetic cultures. Usnic acid, a common constituent of the natural lichen, was not detected in any mycobiont culture or in 16 different synthetic cultures. In contrast, barbatic acid was found in all of the syntheses but in none of the mycobiont cultures. Analyses of synthetic squamules (on mica) and mycobiont strains (from Lilly and Barnett medium) for secondary products gave the following results:

1. *C. cristatella* ss 6 × *T. erici* (inoc. Feb. 11, 1980). CFC 7055. TLC (2308) showed barbatic acid (+ + +), 4-O-dementhylbarbatic acid (+), obtusatic acid (trace), 3-α-hydroxybarbatic acid (+), probably a faint trace of didymic acid, and rhodocladonic acid along with UV (longwave)+ and sterollike spots. HPLC (Aug. 20, 1980, #5–8) showed barbatic acid with traces of didymic, obtusatic, 4-O-demethylbarbatic, and norobtusatic acids. The mycobiont alone did not produce any characteristic lichen substances in culture.

2. *C. cristatella* ss 11 × *T. erici* (inoc. Feb. 15, 1980). CFC 7057. TLC (2308) showed barbatic acid (+ + +), didymic acid (+), higher homologue of didymic acid (+), probable faint trace of 4-O-demethylbarbatic acid, and UV (longwave)+ and sterollike spots. HPLC (Aug. 21, 1980, #6–7) of a well-developed squamule showed barbatic acid, a moderate proportion of didymic acid, the higher homologue of didymic acid, the lower homologue of didymic acid (trace), obtusatic acid (trace), 4-O-demethylbarbatic acid (trace), and norobtusatic acid (trace). The mycobiont alone did not produce any characteristic lichen substances in culture.

3. *C. cristatella* ss 28 × *T. erici* (inoc. Feb. 20, 1980). CFC 7058. TLC (2308) showed barbatic acid (+ +), didymic acid (+ + +), higher homologue of didymic acid (trace), 4-O-demethylbarbatic acid (trace), and UV (longwave)+ and sterollike spots. HPLC (Aug. 21, 1980, #8–10) of a well-developed squamule showed barbatic acid, a high proportion of didymic acid, the higher homologue of didymic acid, obtusatic acid (trace), and 4-O-demethylbarbatic acid (trace). The mycobiont alone produced didymic acid in culture.

4. *C. cristatella* ss 6 × *Friedmannia israeliensis* (inoc. May 8, 1980).

CFC 7056. TLC (2308) showed barbatic acid, some UV (longwave) + spots, and some sterollike spots. HPLC (Aug. 21, 1980, #2–3) showed barbatic and traces of obtusatic acid and 4-O-demethylbarbatic acids. The ability of the mycobiont to produce lichen substances with this alga, which is not a symbiont, indicates that a functional relationship was established in the synthesis. It also shows that whatever the alga contributes to the symbiosis to stimulate the production of lichen substances by the mycobiont may be more generally available than supposed. Culberson and Ahmadjian (1980) postulated that production of lichen products may involve algal inhibition of the fungal enzyme orsellinic acid decarboxylase. Such an inhibitor would be a simple molecule that might be present in different algae.

Effect of humidity on squamule development

In one synthesis series, saturated solutions of different salts (Winston & Bates, 1960) were used to maintain the following levels of humidity (at 20° C) in 125-ml screw-cap Erlenmeyer flasks: 98% (K_2SO_4), 93% (KNO_3), 85% (KCl), 76% (NaCl), 65.5% ($NaNo_2$), and 48.5% (KNO_2).

Another synthesis series was tested using different mixtures of glycerol and water (Braun & Braun, 1958) to achieve the following levels of humidity in the closed flasks: 100%, 95%, 90%, 86%, 80%, 74%, 70%, and 60%. The specific gravities of the solutions were determined with a hydrometer, and the different levels of humidity were verified by means of a wet-dry bulb psychrometer.

In general, the symbionts did poorly in flasks where salt solutions were used to control humidity. The algae turned brown and there was at best only limited growth of the fungus. Only at the 98% level (K_2SO_4) was there a limited presquamule development in two of the five synthesis flasks after 1 month. The mixture of symbionts inoculated into flasks with humidity 85% or lower dried fully in about 18 h.

In the glycerol-water series, aerial growth of the fungus and some presquamule formation were evident in flasks at the 90% and above humidity levels after 2 months. In flasks at 86% humidity and below, the algal cells had died and aerial hyphae of the fungus were not present. Bertsch (1966a) found that photosynthesis in lichens occurred only at relative humidities higher than 80%–85%. Luminescence, another possible measurement of photosynthesis, was especially strong between 95% and 100% relative humidity for the lichen *Cladonia impexa* (Sigfridsson, 1980).

Polysymbiosis and alternate wetting and drying

The idea that more than two symbionts are involved in lichens is still considered viable (Hale, 1974). *Azotobacter*, a nitrogen-fixing bacterium, is generally thought to be the most likely third symbiont. Polysymbiosis can be ruled out in species we have synthesized axenically, at least in terms of morphogenetic development of the thallus. Whether secondary symbionts play a role in the nutritional needs of a lichen is unknown. Bacteria are common on and within a lichen thallus and they may break down organic substances absorbed by lichens from rainwater.

Alternate drying and wetting, thought to be a prerequisite for lichen synthesis, was not necessary in our cultures. After the initial drying of the symbionts on the mica strip, further wetting of the culture did not occur. The only water available to the symbionts was from vapor in the air of the flask. Lange et al. (1970a, b) showed that lichens can absorb water vapor from a nonsaturated atmosphere at a rate sufficient to support a net photosynthetic gain. Nonlichenized aerial algae also can absorb water vapor and grow (Bertsch, 1966b). Fruticose and foliose lichens in an atmosphere of 95% relative humidity can absorb up to 30%–40% of the amount of water they can hold when saturated with liquid water (Blum, 1973). In our humidity series, synthesis did not occur in flasks with a relative humidity of less than 90%. The high relative humidity of our synthesis flasks stimulated growth of aerial hyphae that grew out of the original compact fungal mass.

Summary

Synthesis studies indicate that some lichen fungi are parasites on algae and that the symbiosis is one of controlled parasitism. Evidence for the involvement of lectins in symbiont recognition in lichens is inconclusive. Lichenization may be determined by algal resistance to fungal attack, such resistance possibly caused by phytoalexins. The modifications that algal symbionts undergo when they are in culture may be related to the development of a presymbiotic state that is a prerequisite of lichenization. Developmental stages of squamule formation in *C. cristatella* are described from axenic cultures. Squamules formed only at relative humidities of 90%–100%. The synthetic lichens produced secondary products found in the natural forms. Alternate drying and wetting was not necessary for synthesis. The axenic syntheses proved that bacteria are not necessary for the morphogenetic development of the lichen. Concentric bodies were not present in hyphae of the axenic squamules.

References

Ahmadjian, V. (1962). Investigations on lichen synthesis. *Amer. J. of Bot.* **49**, 277–83.

– (1966). Artificial reestablishment of the lichen *Cladonia cristatella. Science* **151**, 199–201.

– (1967). *The Lichen Symbiosis.* Waltham, Mass.: Blaisdell Publishing Co.

– (1970). The lichen symbiosis: Its origin and evolution. *In: Evolutionary Biology*, vol. 4, ed. Th. Dobzhansky, M. K. Hecht, & W. C. Steere, pp. 163–84. New York: Appleton-Century-Crofts.

Ahmadjian, V., & Jacobs, J. B. (1981). Relationship between fungus and alga in the lichen *Cladonia cristatella* Tuck. *Nature* **289**, 169–72.

Ahmadjian, V., Jacobs, J. B., & Russell, L. A. (1978). Scanning electron microscope study of early lichen synthesis. *Science* **200**, 1062–4.

Ahmadjian, V., Russell, L. A., and Hildreth, K. C. (1980). Artificial reestablishment of lichens. I. Morphological interactions between the phycobionts of different lichens and the mycobionts *Cladonia cristatella* and *Lecanora chrysoleuca*. *Mycologia* **72**, 73–89.

Albersheim, P., and Anderson-Prouty, A. J. (1975). Carbohydrates, proteins, cell surfaces, and the biochemistry of pathogenesis. *Ann. Rev. of Plant Phys.* **26**, 31–52.

Albersheim, P., & Valent, B. S. (1978). Host-pathogen interactions in plants. *J. Cell Biol.* **78**, 627–43.

Archibald, P. A. (1975). *Trebouxia* de Puymaly (Chlorophyceae, Chlorococcales) and *Pseudotrebouxia* gen. nov. (Chlorophyceae, Chlorosarcinales). *Phycologia* **14**, 125–37.

Ascaso, C., & Galvan, J. (1975). Concentric bodies in three lichen species. *Arch. Microbiol.* **105**, 129–30.

Barrett, J., & Howe, M. L. (1968). Hemagglutination and hemolysis by lichen extracts. *App. Microbiol.* **16**, 1137–9.

Bartusch, H. (1931). Beiträge zur Kenntnis der Lebensgeschichte des Xanthoriapilzes. *Arch. für Mikrobiologie* **3**, 122–57.

Bertsch, A. (1966a). Über den CO_2-Gaswechsel einiger Flechten nach Wasserdampfaufnahme. *Planta* **68**, 157–66.

– (1966b). CO_2-Gaswechsel und Wasserhaushalt der Aerophilen Grünalgen *Apatococcus lobatus*. *Planta* **70**, 46–72.

Blum, O. B. (1973). Water relations. *In: The Lichens*, ed. V. Ahmadjian & M. E. Hale, pp. 381–400. New York: Academic Press.

Bonnier, G. (1889). Recherches sur la synthèse des lichens. *Ann. Sci. Nat.* Série 7. **9**, 1–34.

Braun, J. V., & Braun, J. D. (1958). A simplified method of preparing solutions of glycerol and water for humidity control. *Corrosion* **14**, 17–18.

Bubrick, P., & Galun, M. (1980). Symbiosis in lichens: Differences in cell wall properties of freshly isolated and cultured phycobionts. *Federation of European Microbiological Societies (FEMS) Microbiology Letters* 7, 311–13.

Bubrick, P., Galun, M., Ben-Yaacov, M., & Frensdorff, A. (1982). Antigenic similarities and differences between symbiotic and cultured phycobionts from the lichen, *Xanthoria parietina*. *FEMS Microbiology Letters* **13**, 435–8.

Chantanachat, S., & Bold, H. C. (1962). Phycological studies. II. Some algae from arid soils. *Univer. of Texas Publi.* No. 6218, 75 pp.

Collins, C. R., & Farrar, J. F. (1978). Structural resistances to mass transfer in the lichen *Xanthoria parietina*. *New Phytol.* **81**, 71–83.

Culberson, C. F., & Ahmadjian, V. (1980). Artificial reestablishment of lichens. II. Secondary products of resynthesized *Cladonia cristatella* and *Lecanora chrysoleuca*. *Mycologia* 72, 90–109.

Deason, T. R., Ryals, P. E., O'Kelley, J. C., & Bullock, K. W. (1979). Fine structure of mitosis and cleavage in *Friedmannia israeliensis* (Chlorophyceae: Chlorsarcinaceae). *J. Phycol.* **15**, 452–7.

DeBary, A. (1866). *Morphologie und Physiologie der Pilze, Flechten, und Myxomyceten*. Leipzig: Engelmann.

Drew, E. A., & Smith, D. C. (1967). Studies in the physiology of lichens. VII. The physiology of the *Nostoc* symbiont of *Peltigera polydactyla* compared with cultured and free-living forms. *New Phytol.* **66**, 379–88.

Durrell, L. W. (1967). An electron microscope study of algal hyphal contact in lichens. *Mycopathologia et Mycologia Applicata* **31**, 273–86.

Elenkin, A. (1902). Zur Frage der Theorie des Endosaprophytismus bei Flechten. *Bull. Jard. Imp. Bot. St. Petersb.* **2**, 65–84.

Farrar, J. F. (1976a). Ecological physiology of the lichen *Hypogymnia physodes*. I. Some effects of constant water saturation. *New Phytol.* **77**, 93–103.

– (1976b). Ecological physiology of the lichen *Hypogymnia physodes*. II. Effects of wetting and drying cycles and the concept of "physiological buffering." *New Phytol.* **77**, 105–13.

Filho, X. L., Mendes, L. C. G., & Vasconcelos, C. A. F. (1971). Fitohemaglutinina em alguns criptógamos. *Universidade Federal de Pernambuco Instituto de Biociencias Departmento de Botanica*, série B, *Estudos e Pesquisad* **2**, 1–8.

Fisher, K. A., & Lang, N. J. (1975). Comparative ultrastructure of cultured species of *Trebouxia*. *J. of Phycol.* **7**, 155–65.

Galun, M., Ben-Shaul, Y., & Paran, N. (1971). The fungus-alga association in the Lecideaceae: An ultrastructural study. *New Phytol.* **70**, 483–5.

Green, T. G. A., & Smith, D. C. (1974). Lichen physiology. XIV. Differences between lichen algae in symbiosis and in isolation. *New Phytol.* **73**, 753–66.

Greenhalgh, G. N., & Anglesea, D. (1979). The distribution of algal cells in lichen thalli. *Lichenologist* **11**, 283–92.

Hale, M. E. (1974). *The Biology of Lichens* 2nd ed. New York: Elsevier.

Heslop-Harrison, J. (1978). *Cellular Recognition Systems in Plants*. Baltimore: University Park Press.

Hessler, R., & Peveling, E. (1978). Die Lokalisation von ^{14}C-Assimilaten in Flechtenthalli von *Cladonia incrassata* Floerke und *Hypogymnia physodes* (L.) Ach. *Zeitschr. Pflanzenphysiol.* **86**, 287–302.

Hildreth, K. C., & Ahmadjian, V. (1981). A study of *Trebouxia* and *Pseudotrebouxia* isolates from different lichens. *Lichenologist* **23**, 65–86.

Hori, T. (1973). Comparative studies of pyrenoid ultrastructure in algae of the *Monostroma* complex. *J. of Phycol.* **9**, 190–9.

Jacobs, J. B., & Ahmadjian, V. (1971). The ultrastructure of lichens. II. *Cladonia cristatella*: The lichen and its isolated symbionts. *J. of Phycol.* **7**, 71–82.

Kershaw, K. A., & Millbank, J. W. (1970). Nitrogen metabolism in lichens. II. The partition of cephalodial-fixed nitrogen between the mycobiont and phycobionts of *Peltigera aphthosa*. *New Phytol.* **69**, 75–9.

Lange, O. L., Schulze, E. D., & Koch, W. (1970a). Experimentellökologische Untersuchungen an Flechten der Negev-Wüste. II. CO2-Gaswechsel und Wasserhaushalt von *Ramalina maciformis* (Del.) Bory am natürlichen Standort während der sommerlichen Trockenperiode. *Flora* **159**, 38–62.

– (1970b). Experimentellökologische Untersuchungen an Flechten der Negev-Wüste. III. CO2-Gaswechsel und Wasserhaushalt von Krusten- und Blattflechten am natürlichen Standort während der sommerlichen Trockenperiode. *Flora* **159**, 525–38.

Lange-de la Camp, M. (1933). Kulturversuche mit Flechtenpilzen (*Xanthoria parietina*). *Arch. für Mikrobiologie* **4**, 379–93.

Lichtenthaler, H. K. (1968). Plastoglobuli and the fine structure of plastids. *Endeavour* **27**, 144–9.

Lockhart, C. M., Rowell, P., & Stewart, W. D. P. (1978). Phytohaemagglutinins from the nitrogen-fixing lichens *Peltigera canina* and *P. polydactyla*. *Federation of European Microbiological Societies (FEMS) Microbiology Letters* **3**, 127–30.

Molnar, K. E., Stewart, K. D., & Mattox, K. R. (1975). Cell division in the filamentous *Pleurastrum* and its comparison with the unicellular *Platymonas* (Chlorophyceae). *J. of Phycol.* 11, 287–96.
Peirce, G. J. (1900). The relation of fungus and alga in lichens. *Amer. Nat.* 34, 245–53.
Petit, P. (1982). Phytolectins from the nitrogen-fixing lichen *Peltigera horizontalis*: The binding pattern of primary protein extract. *New Phytol.* 91, 705–10.
Peveling, E. (1973). Fine structure. *In: The Lichens*, ed. V. Ahmadjian & M. E. Hale, pp. 147–82. New York: Academic Press.
Plessl, A. (1963). Über die Beziehungen von Haustorientypus und Organisationshohe bei Flechten. *Österreichischen Botanischen Zeitschrift* 110, 194–269.
Puymaly, A. de (1967). Ce que l'on désigne par symbiose chez les Lichens n'est-ce pas plutôt du domaine du parasitisme? *Bull. Soc. bot. Fr., Colloque sur les Lichens*, 121–7.
Quispel, A. (1943–5). The mutual relationships between algae and fungi in lichens. *Recueil des Travaux Botaniques Néerlandais* 40, 413–541.
Richardson, D. H. S. (1973). Photosynthesis and carbohydrate movement. *In: The Lichens*, ed. V. Ahmadjian & M. E. Hale, pp. 249–88. New York: Academic Press.
Schwendener, S. (1868). Ueber die Beziehungen zwischen Algen und Flechtengonidien. *Botanische Zeitung* 26, 289–92.
Sigfridsson, B. (1980). Some effects of humidity on the light reaction of photosynthesis in the lichens *Cladonia impexa* and *Collema flaccidium*. *Physiol. Plant.* 49, 320–6.
Silverberg, B. A., & Sawa, T. (1974). An ultrastructural study of the pyrenoid in cultured cells of *Chlorella variegata* (Chlorococcales). *New Phytol.* 73, 143–6.
Slocum, R. D., Ahmadjian, V., & Hildreth, K. C. (1980). Zoosporogenesis in *Trebouxia gelatinosa*: Ultrastructure, potential for zoospore release, and implications for the lichen association. *Lichenologist* 12, 173–87.
Smith, D. C. (1967). The movement of carbohydrate from alga to fungus in lichens. *Bull. Soc. Bot. Fr., Colloque sur les Lichens*, 129–33.
– (1980). Mechanisms of nutrient movement between the lichen symbionts. *In: Cellular Interactions in Symbiosis and Parasitism*, ed. C. B. Cook, P. W. Pappas, & E. D. Rudolph, pp. 197–227. Columbus: Ohio State University Press.
Thomas, E. A. (1939). Über die Biologie von Flechtenbildnern. *Beitr. Kryptogamenflora Schweiz* 9, 1–208.
Winston, P. W., & Bates, D. H. (1960). Saturated solutions for the control of humidity in biological research. *Ecology* 41, 232–7.

9

Subaerial symbiotic green algae: interactions with vascular plant hosts

RUSSELL L. CHAPMAN
Department of Botany
Louisiana State University
Baton Rouge, LA 70803

BARRY H. GOOD
Department of Biological Sciences
Loyola University
New Orleans, LA 70118

The Chroolepidaceae for the most part are not well known, even among professional phycologists. Nevertheless, a discussion of the interaction between subaerial symbiotic green algae and vascular plant hosts must concentrate on this unusual family. Thus, this chapter is devoted almost entirely to the Chroolepidaceae, although a brief review of some pertinent information on other green algae has been included. The best-known genus in the family is *Trentepohlia*, and the family is frequently referred to as the Trentepohliaceae. However, Papenfuss (1962) noted that "Chroolepidaceae" (Rabenhorst 1868) has priority over "Trentepohliaceae" (Hansgirg 1886). The Chroolepidaceae include the following genera: *Cephaleuros*, *Phycopeltis*, *Stomatochroon*, and *Trentepohlia*. The four genera are distinguished by differences in morphology and, to some extent, the mode of interaction with vascular plant hosts. The genus *Physolinum* was included in the family by Printz (1939, 1964); however, Flint (1959) returned the genus to *Trentepohlia*. Also, the late Dr. Rufus H. Thompson created a new genus, *Printzia*, which he described in an as yet unpublished manuscript. (This monograph is currently being prepared for publication by Dr. Peter Timpano and is hereafter cited as "Thompson and Timpano, personal communication.") Some authors (e.g., Smith 1950; Fritsch 1965; Prescott 1968) assign the aquatic taxa *Ctenocladus*, *Gomontia*, and *Gongrosira* to the Chroolepidaceae. Retention of these taxa in the Chroolepidaceae is no longer tenable (see e.g., Blinn & Morrison 1974; Chapman & Good 1978; Chappel et al. 1978). All

The authors wish to acknowledge with gratitude the expert technical assistance and general helpful suggestions of Margaret C. Henk and those who kindly provided photographs (Judy Meier, Department of Botany, Louisiana State University, Fig. 21; Robert B. Marlatt, University of Florida, Gainesville, Fig. 8). We thank the official and numerous unofficial reviewers for their most helpful comments and suggestions and wish to acknowledge with special gratitude the gracious assistance of the late Dr. Rufus H. Thompson, Department of Botany, University of Kansas. Some of the research reported in this chapter was supported in part by NIH Biomedical Research Support Grant SO7 RR07039-07.

members of this family are filamentous and subaerial, some occurring on inanimate substrates such as rock and wood, others on vascular plant hosts. Although especially abundant in tropical and subtropical regions of the world, some Chroolepidaceae do occur in more temperate climatic regions. The following are among the key characteristics of the family. First, there are cytoplasmic accumulations of the yellow to orange-colored pigment hematochrome. Second, the cell walls tend to be especially thick compared to those of many aquatic green algae. Third, the zoosporangia abscise and are dispersed by wind, rain, insects, and arachnids. Fourth, some ultrastructural, cytological, and biochemical features (to be discussed later in the chapter) are characteristic of this family.

An additional interesting feature of the Chroolepidaceae is that at least some of the genera combine with fungi to form lichens (see Chapter 8 in this volume). Thus, these algae are among the few filamentous green algae known to function as phycobionts. Although the alga-fungus relationship is not a topic of interest per se for this chapter, the state of the algae (i.e., free-living vis-à-vis phycobiont) does affect the interaction with vascular plant hosts. Therefore, some discussion of the lichenized states will be included in the section on the genus *Cephaleuros*, which is the only genus for which such modified interaction with the vascular plant host is a pertinent topic. In fact, most of the chapter will be devoted to the genus *Cephaleuros* for two reasons. First, the extent of interaction with the vascular plant host is most easily visible and most clearly documented for this genus. Second, very little information is available about the interaction of the other genera with their vascular plant hosts. The rather detailed presentation on *Cephaleuros* will serve as a basis for comparison for later discussion on the other taxa and their interactions, if any, with vascular plant hosts. The chapter will conclude with some brief notes on other green algae that interact with vascular plant hosts.

The genus *Cephaleuros*

Cephaleuros occurs in tropical and subtropical regions and is an epiphyte or parasite of numerous species of vascular plant hosts. For example, Holcomb (1975) reported finding *C. virescens* Kunze on more than 115 species of hosts in Louisiana, and Batista and Lima (1949) cited 448 hosts in Brazil. A list of all host species worldwide would be expected to be even longer. The alga can grow on leaves, young stems, and fruits and is considered to be a pathogen of crop plants such as coffee, tea, and citrus (see e.g., Swingle 1894; Went 1895; Mann & Hutchinson 1907; Winston 1938; Wellman 1965, 1972; Golato 1970; Roth 1971; Joubert & Rijkenberg 1971a; Vidhyasekaran & Parambaramani 1971a, b). The terms *red rust* and *algal rust* have been applied to *Cephaleuros* infections and have caused some confusion by connotating fungal diseases. Al-

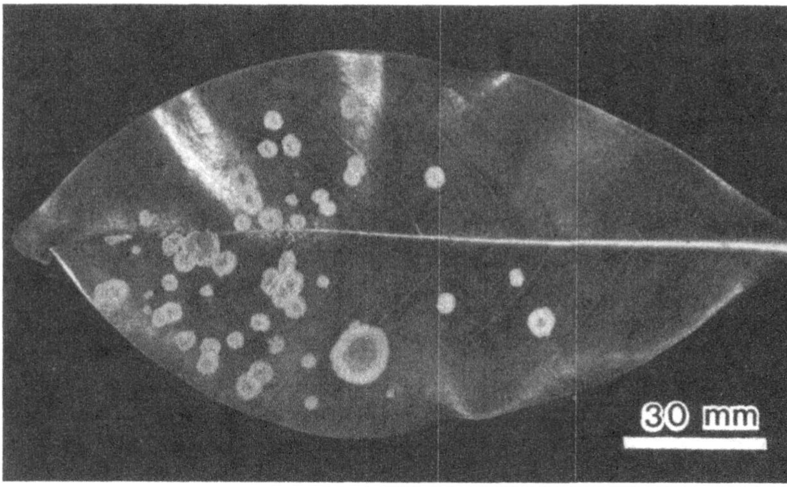

Fig. 1. *Cephaleuros* on leaf of *Magnolia grandiflora*.

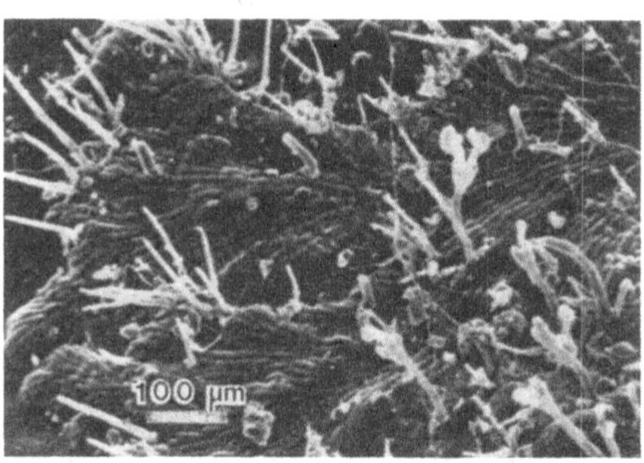

Fig. 2. *Cephaleuros* thallus. Scanning electron micrograph.

though the topic of host specificity will be discussed later in this chapter, it should be noted that infection is not limited to one or a few families of vascular plants.

Very young vegetative thalli (consisting of only a few cells) of *Cephaleuros* are microscopic. Such thalli bear few erect branches and contain little visible hematochrome pigment. More mature thalli are clearly visible to the naked eye (Fig. 1) and are often bright orange. During periods of active asexual reproduction, the thalli are velvety in appearance and may cover large areas of the host surface in the area of infection. In such instances numerous thalli may abut and in some cases overlap.

The velvety appearance of mature thalli is created by abundant setae (also called sterile trichomes, hairs, or filaments) and fertile branches (Fig. 2). Although these structures clearly extend above the host cuticle,

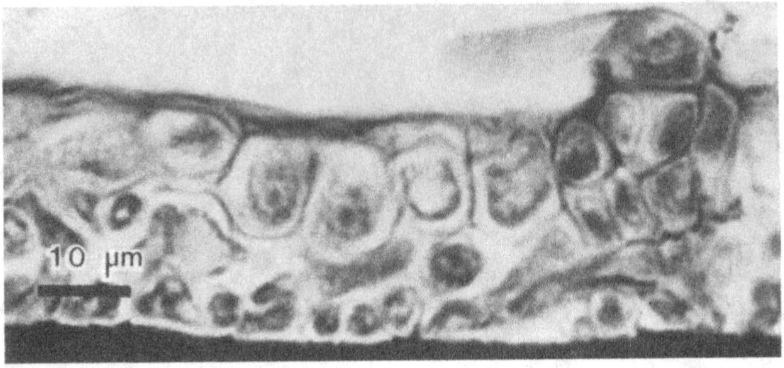

Fig. 3. *Cephaleuros* thallus filaments beneath host cuticle. Note erect filament. Light micrograph.

the prostrate thallus from which they arise is subcuticular; thus the erect branches have erupted through the cuticle (Fig. 3). The prostate thallus is composed of compacted branched filaments that form recognizable radiating lobes or "ramuli." Beneath the prostrate thallus ventral branches or rhizoids are formed. In some cases (e.g., *C. virescens* infections of *Magnolia grandiflora* Linnaeus leaves; Chapman, 1976b), the rhizoids grow beneath the prostrate thallus and above the host epidermal cells; in others (e.g., *C. parasiticus* Karsten), filaments may grow through intercellular spaces into necrotic host tissue subtending the algal thallus.

Ultrastructural studies (including Rijkenberg et al. 1971; Chapman 1976a; Marche-Marchad 1976, 1977; Chapman & Good 1978; Chappel et al. 1978) have demonstrated the presence of simple plasmodesmata in the crosswalls of *Cephaleuros* vegetative and reproductive cells. In the absence of experimental data, one is free to speculate on the functional significance of the intercellular connections provided by the plasmodesmata in this alga. Although the alga is heterotrichous, all the cells are photosynthetic and translocation of photosynthates may not be very important. Since some portions of the thallus (viz. the trichomes and fertile branches) are exposed supracuticularly, efficient adjustment of water potential may be necessary. Plasmodesmata may facilitate transfer as well as translocation of host-derived organic and/or inorganic compounds. Hypotheses of plasmodesmata function in *Cephaleuros* should include consideration of the role of similar plasmodesmata in *Phycopeltis* and *Trentepohlia*, which are different in morphology and habitat.

Starch grains (Fig. 4) occur within the chloroplasts and although some authors (e.g., Bourrelly 1966) state that starch does not occur in the Chroolepidaceae, recent enzymatic studies (Chapman, unpublished results) clearly indicate that the grains are composed of true (i.e., amylose-rich) starch rather than amylopectin or glycogen. Some unusual carbo-

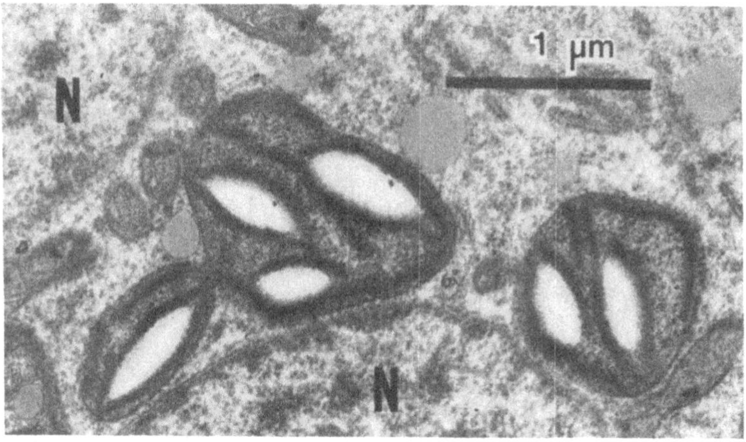

Fig. 4. Starch grains within chloroplast of *Cephaleuros*. N, nucleus. Transmission electron micrograph.

hydrates are produced and accumulated in *Cephaleuros* and *Trentepohlia* (Feige & Kremer 1980). In fact, the pattern of polyhydroxy alcohols (polyols and alditols) in this family is considered "a unique feature as compared with the chemical characters of the majority of other green algae so far investigated" (Feige & Kremer 1980). Although there appears to be a correlation between the accumulation of more than one polyhydroxy alcohol (and, in some instances, the concomitant absence of sucrose) and subaerial growth, the ecological significance, if any, has not been studied.

Asexual reproduction involves many structures and features that clearly facilitate the subaerial existence of *Cephaleuros* and that are unique to the Chroolepidaceae. A discussion of asexual reproduction must begin with definitions of terms. For the most part, the terminology preferred by Thompson (personal communication) will be used; following the initial appearance of each term, synonyms will be listed parenthetically. The structures illustrated in Fig. 5 are *fertile branches* arising from the prostrate thallus. The *sporangiophore* is an erect multicellular filament that terminates in the bulbous *head cell* (pyriform cell, *Kopfzelle*). The head cell bears one to eight (or more) *suffultory cells* (pedicels, stalk cells, neck cells, sporangium mother cells), each of which produces, terminally, one zoosporangium (*Hackensporangien*). A single papillate escape pore is formed on each zoosporangium. The suffultory cell and its terminal zoosporangium are termed a *sporangiate-lateral*.

The fertile branch morphology facilitates dispersal of *Cephaleuros* by elevating zoosporangia above the host cuticle and the prostrate thallus, where they are more easily subject to dispersion by wind, rain, arachnids

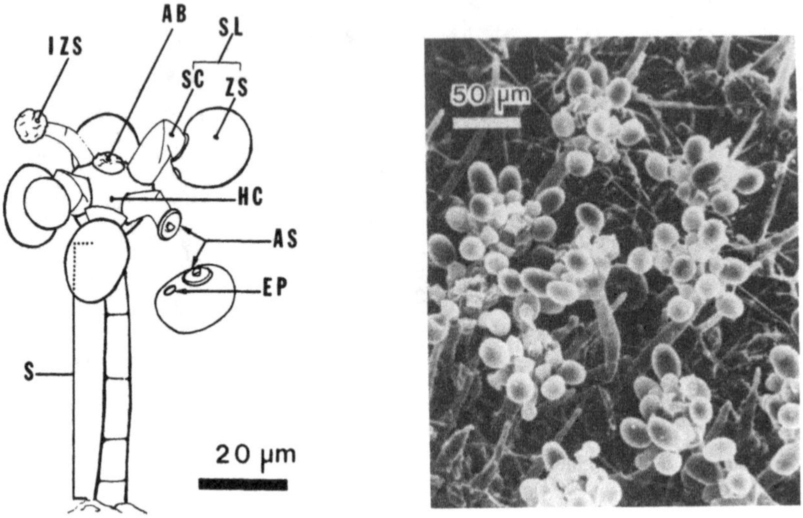

Fig. 5. Left. Fertile branch of *Cephaleuros.* S, sporangiophore; IZS, immature zoosporangium; AB, apical bulge; SL, sporangiate lateral; SC, suffultory cell; ZS, zoosporangium; HC, head cell; AS, abscission septum; EP, escape pore. **Right.** *Cephaleuros* fertile branches. Scanning electron micrograph.

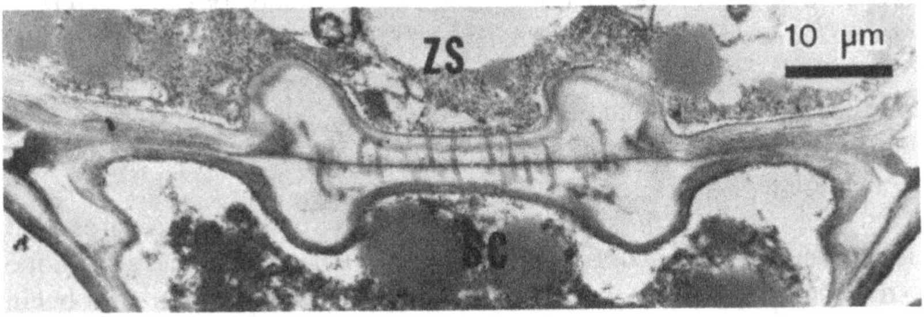

Fig. 6. Abscission septum. Note central ring of wall material, localized plasmodesmata. ZS, zoosporangium; SC, suffultory cell.

(spiders and mites), and insects. Although dispersal of the zoosporangia is passive, the actual abscission of the zoosporangia is preceded by a series of morphological changes and is a feature unique to the Chroolepidaceae. The abscission process has been described for *Cephaleuros* (Chapman 1976b) and *Phycopeltis* (Good & Chapman 1978b) and can be summarized as follows: The specialized crosswall between the zoosporangium and the suffultory cell is the "abscission septum" (also termed the zoosporangium pedicel septum). There is a central area containing simple plasmodesmata surrounded by a ring of thickened wall material (Fig. 6). The perimeter of the abscission septum is a second thickened ring of wall material. The crosswall between the inner ring and outer ring is laminate and contains no plasmodesmata. Enlargement of the zoosporangium, together with turgor changes in the suffultory cell, ultimately results in a circumscissile splitting of the outer ring of wall material in the abscission septum. Thus, the zoosporangium remains attached by the central ring only. As mentioned, the suffultory cell is turgid or flaccid in direct response to the cell's water potential. When conditions are dry (hence unfavorable for the final release of water-dependent zoospores), the suffultory cells are flaccid and pull the zoosporangia close to the head cell. When turgid, the suffultory cells hold the zoosporangia out and away from the head cell. Also, the portion of the abscission septum in the suffultory cell bulges outward, thereby further straining the remaining connection to the zoosporangium.

Concomitant with enlargement of the zoosporangium, cytoplasmic maturation of the zoospores occurs. Repeated mitoses produce 8, 16, or more nuclei and progressive cytoplasmic cleavage forms 8, 16, or more uninucleate, quadriflagellate zoospores. Early in this ontogeny an escape pore forms in the zoosporangium wall. The wall remains thin at the site of the pore and a plug of pectinaceous material develops as the remainder of the zoosporangium wall thickens. The plug dissolves and allows the release of the zoospores whenever the zoospores are sufficiently mature and adequate moisture is available. Zoospore release is not dependent on abscission of the zoosporangium. As may be inferred from the preceding description, the mature zoosporangia are not a form of akinete or resistant stage and have not been shown to be a means of long-term or long-distance dispersal.

Just as the zoosporangia do not provide a means for extensive dispersal, the zoospores cannot exist long after their release. They must settle quickly and begin forming a new thallus. If they are located on an inappropriate substrate or host, they will perish. Since all *Cephaleuros* species grow beneath the host cuticle, either the zoospores or the young thalli must actively penetrate the cuticle or fortuitously settle in a scratch, wound, or other break in the cuticle. Active penetration by the small, short-lived zoospore seems unlikely, and there is no evidence to support the hypothesis that young thalli dissolve subtending cuticular waxes. The hypothesis that infection is predicated on preexisting breaks in the cuticle would account for the absence of infections on young leaves, which presumably have fewer scratches and nicks in the cuticle. As will be discussed, the initiation of infection is host specific and understanding the mode of entry to a subcuticular site is prerequisite to understanding the basis of host specificity. Therefore, the possible role of insects and arachnids in both dispersal and infection should be noted.

Both insects and arachnids have been observed to transport *Cephaleuros* zoosporangia (Thompson & Timpano, personal communication;

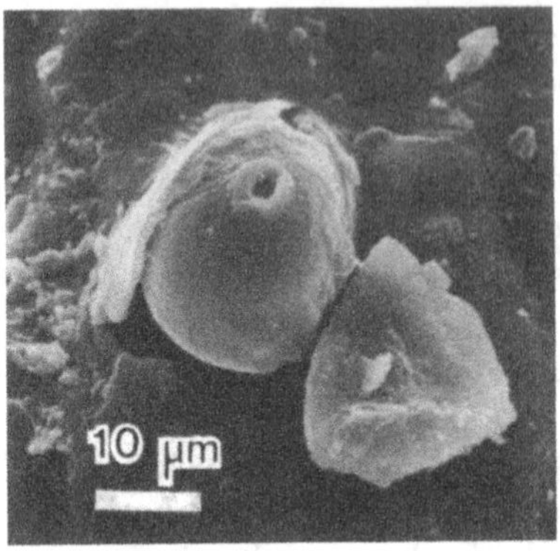

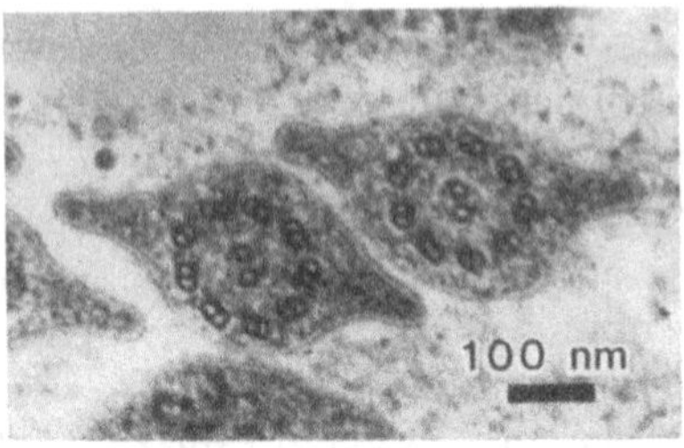

Fig. 7. Top left. *Cephaleuros* zoospore. Note flat flagella. Scanning electron micrograph. **Top right.** Keeled flagella of *Cephaleuros* zoospore as seen in cross section. Transmission electron micrograph. **Bottom left.** *Cephaleuros* gametangium. Note escape pore, broken host cuticle, and outline of the subcuticular filaments. Scanning electron micrograph.

Henk & Chapman, unpublished observations). These agents of dispersal also feed on the leaves of the host plants. Thus, holes or tears that penetrate the cuticle are produced on suitable hosts by organisms that may bear the algal zoosporangia. A *Cephaleuros* infection may be facilitated further by the carbohydrate-containing exudate that usually remains in the feeding sites. A positive correlation between number of infections (i.e., algal thalli) per leaf and the number of insect galls (Chowdary & Jose 1979) implies an insect role in infection on *Mangifera indica* Linnaeus.

The zoospores are certainly a significant means of propagation, if not the most important. These cells are also significant factors in the delineation of the algal family. The presence of keeled (or winged or flanged) flagella (Fig. 7, top left and right) is a unique feature of these taxa. First reported for gametes of *Trentepohlia* (Graham & McBride 1975), keeled flagella are the only type reported for gametes and zoospores of Chroolepidaceae that have been studied. Another unusual feature of the flagella is the twisting and close adherence of paired flagella of quadriflagellate zoospores. This phenomenon has created some confusion because quadriflagellate zoospores often appear to be biflagellate. Multilayered structures associated with the basal bodies and "reversed bilateral symmetry" (Chapman 1980, 1981) are also important phylogenetic characters.

Sexual reproduction is a second mode of propagation and host infection. Generally, gametangia are formed by terminal cells of the prostrate filaments and thus are located beneath the host cuticle. Enlargement of the gametangia forces them upward through the cuticle, eventually exposing the exit pores (Fig. 7, bottom left). The gametangial initials are uninucleate and undergo a series of mitoses and progressive cytoplasmic cleavages similar to those in zoosporangia. The number of gametes formed is 8 to 64 or more (cf. 8–16 or more zoospores) and they are biflagellate. Although some investigators report that the gametes can develop into new thalli without fusion with other gametes (thus paralleling reproduction via zoospores), Thompson (1959) has provided an account of the complete sexual cycle. Gametes fuse either within the gametangium (usually when conditions are dry) or external to it. The zygote immediately initiates development of a dwarf sporophyte, which in turn produces small, presumptive meiozoosporangia. These meiozoosporangia are abscised and release four to eight quadriflagellate meiozoospores, which presumably infect a host and produce the typical vegetative thallus. It should be noted that the dwarf sporophytes are formed quite near the parent thallus, often in or near the gametangia themselves.

A third mode of propagation is the formation of akinetes, which are resistant reproductive cells. Akinete production is triggered by dry conditions and constitutes a second mode of development for the sporangiophore initials (i.e., those cells in the prostrate, subcuticular filaments that normally produce the fertile branches). The sporangiophore initials develop thick walls and enlarge. When environmental conditions favor further development, the akinetes produce either a fertile branch or 32 quadriflagellate zoospores (i.e., quadriflagellate cells that are indistinguishable from the zoospores produced in regular zoosporangia). This latter development has confused some observers because the situation resembles gametangial development; however, gametangia release biflagellate cells, not quadriflagellate cells.

As a final point in this discussion of life cycle and propagation, it should be noted that both sexual and asexual reproduction continues over many months (e.g., in Louisiana one or both forms of reproduction occur during no fewer than 6 months of each year). Further, each thallus produces hundreds of fertile branches, each of which bears four or more zoosporangia. Each zoosporangium produces 8 to 16 or more zoospores. Thus each thallus produces thousands of zoospores that can infect or reinfect hosts. Similarly, a large number of meiozoospores can be produced as well. The fact that an infected host is not usually completely covered with *Cephaleuros* may indicate that very few zoospores or meiozoospores successfully produce new thalli.

Hosts: specificity, responses, and interaction

The subject of host specificity entails an interesting paradox. *Cephaleuros* (especially *C. virescens*) occurs on so many different host species representing so many families of vascular plants that clearly there is no restriction to particular taxonomic groups, nor is infection limited to a single leaf type. (The observation that infection is restricted to coriaceous leaves has been reported by some authors, but should be considered a limited observation based on a relatively small number of examples.) The apparent lack of uniformity among susceptible hosts can lead to the suggestion that there is *no* specificity, but quite clearly *Cephaleuros* does not infect any and all vascular plants. It is not uncommon to find infected plants side by side with uninfected plants and such situations continue for many years. Wellman (1972) cites examples of clear differences in susceptibility between different species of the same genus as well as between different varieties of the same species. Marlatt and Campbell (1980) reported statistically different responses to *Cephaleuros* infections in cultivars of guava. There must be certain prerequisites all susceptible hosts meet, but these are simply not obvious. Young, rapidly expanding, and undamaged leaves are not generally infected, even on plants known to be susceptible. Similarly, short-lived leaves such as those of most annuals are not generally infected. Broad-leafed evergreens provide substrates that are available for longer-term infection and successive reproductive cycles, but other factors affecting susceptibility must also be involved. If infection is predicated on attachment to and penetration through the host cuticle, the chemical nature, surface topography, and thickness of this waxy layer could all affect susceptibility. If infection requires preexisting breaks in the cuticle, the chemical nature and thickness of the cuticle could still be important. The possible role of insects and arachnids as both vectors of dispersal and agents of infection further complicates the question of susceptibility. Beyond the initiation of infection, successful growth of the alga could be affected by the nature of the host response. Such responses might well vary among related taxa and thereby explain the susceptibility of one and resistance of another. The topic of host specificity has not been well studied and, accordingly, many basic questions await experimental investigation. It should perhaps also be noted that even descriptive study (i.e., an extensive listing of all *Cephaleuros* species and all hosts for each species) is far from complete. Given that these algae are widely distributed in the tropics and subtropics, it is understandable that such a study has not yet been made and doubtful that such a study will be undertaken before many hosts and habitats are destroyed by current "development" of tropical rain forests.

The fact that *C. virescens* is the most widely collected species raises the question: Is it the least host specific? The limited data available support this hypothesis, and it would be interesting to know if biochemical and cytological data would evidence a more "ancestral" evolutionary position for *C. virescens* vis-à-vis the 12 other species.

Just as the susceptibility of a vascular plant host to infection by *Ce-*

phaleuros may entail several factors, so also do the host responses. At the onset, a discussion of host responses must recognize four major factors. First, there are 13 species of *Cephaleuros* (Thompson & Timpano, personal communication) and these species should not be expected to induce entirely similar responses in a given host (assuming, of course, that a given host might be susceptible to all 13 species). Second, a given species of *Cephaleuros* can in some instances (viz. *C. virescens*) infect hundreds of different host species, and there is little reason to expect that the host response would be very similar in so many and varied hosts. In addition to the differences among hosts, differences in the condition of a given host species can be expected to affect the nature of the host response. A host plant weakened by a lack of nutrients, by a fungal pathogen, or by any of numerous other factors would probably evidence a more severe reaction to a *Cephaleuros* infection than would a vigorously growing, uninfected plant. A third major factor to be considered in a discussion of host responses is the site of infection. Since *Cephaleuros* can attack leaves, young stems, and fruits, the host response can be quite different in the different organs. Finally, the rather broad topic of environmental factors must be considered a fourth major parameter. An examination of an infection by a single species of *Cephaleuros* restricted to the leaves (rather than leaves and other organs) of a single host species could reveal differences in the host response to infections initiated during different seasons of the year. Although such differences might be more dramatic in the semitropical areas of the distribution range (where seasonal changes are greater) than in the tropical regions, the vigor of the alga, and hence the nature of the host response, would vary in relation to the drier and wetter seasons in tropical areas.

The combination of 13 species of *Cephaleuros*, numerous hosts, and multiple factors that might affect the host responses provides ample possibility for comparative studies. Although some comparative studies have been reported (see Chapman 1976b), especially with respect to plant pathology studies, no comprehensive comparative study has been undertaken. It may be of interest to note that both the differences in algal morphology and in host responses to algal infections recorded for species of *Cephaleuros* have been questioned from time to time because no comparative data were available. Several species of *Cephaleuros* were only known from infections of a single host; hence it was not clear whether the differences in algal morphologies were valid species differences or merely different phenotypic expressions elicited by different hosts. The need for comparative data obtained experimentally through cross-inoculations has long been recognized, but very few, limited studies have been reported. The absence of answers to numerous basic questions concerning the host responses to infection does not belittle the many questions that will appear in the following descriptions of host responses.

Fig. 8. Guava fruits with *Cephaleuros* spots (thalli).

A list of some ornamental and crop plants susceptible to *Cephaleuros* infections includes:

Avocado (*Persea americana* Miller)
Cacao (*Theobroma cacao* Linnaeus)
Citrus spp. (including grapefruit, lime, and orange)
Coffee (e.g., *Coffea arabica* Linnaeus)
Guava (*Psidium guajava* Linnaeus)
Para rubber tree (*Hevea brasiliensis* Mueller)
Magnolia spp.
Mango (*Mangifera indica* Linnaeus)
Oil palm (*Elaeis guineensis* Jacquin)
Pecan (*Carya illinoensis* Koch)
Persimmon (*Diospyros kaki* Linnaeus)
Tea (*Thea sinensis* Linnaeus)
Kumquat (*Fortunella margarita* Swingle)
Camellia spp.

Many of the plants infected by *Cephaleuros* are crop or ornamental plants and therein lies the basis for much of the interest in and study of the genus. Nevertheless, Koch's postulates have not yet been fulfilled and it cannot be stated unequivocally that *Cephaleuros* is the sole causative agent of diseases heretofore ascribed to the alga. Rufus H. Thompson believed that the alga was significantly deleterious in very few instances and that severe infections were opportunistic growths of the alga in diseased or otherwise stressed plants.

As mentioned earlier, the nature of a host response to *Cephaleuros* infection varies with several factors. Infections of fruit in plants such as *Citrus* spp., avocado, guava, and coffee berries directly affect the marketable plant product. Nevertheless, the actual damage is often merely the presence of unattractive, superficial spots (the algal thalli) rather than significant destruction of tissue or perturbation of normal fruit development (Fig. 8). As in the case of leaf infections (discussed later in this

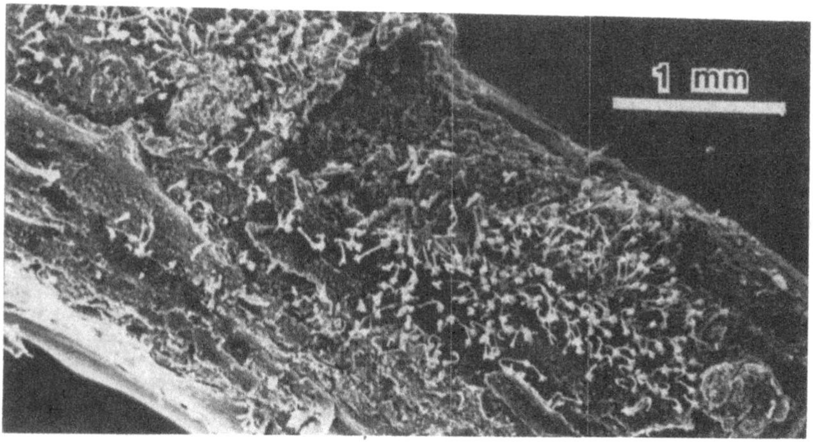

Fig. 9. *Cephaleuros* on a young twig. Scanning electron micrograph.

section), there may be a wound response (viz. cell death) in the cell layers directly beneath the algal thalli.

The infection of young stems (Fig. 9) is more serious than infections of fruits, and partial or complete girdling can occur. Impaired translocation results in poor growth manifested in stunting, fewer (and sometimes abnormal) leaves, and death of the plant. Host stem cells subtending the alga die and the net effect of the infection will depend on the rate of algal growth vis-à-vis the rate of tissue formation in the host. Thus, as expected, very young rapidly growing stems are not seriously infected, and old stems bearing a thick bark are not infected at all.

The infection of leaves by *Cephaleuros* may produce an unsightly condition, but unless the infection is extremely severe, that is, covering a large percentage of the leaf surface, there is little or no significant harm to the plant. Unseasonal abscission of leaves in response to severe infection can be an economically important problem. The host response to the alga varies from no detectable anatomical response in subtending tissues to necrosis extending from surface to surface through the leaf. In leaves that exhibit a wound response (e.g., *M. grandiflora*, Fig. 10), there can be some alga-induced mitotic activity in the epidermis and palisade parenchyma. Joubert and Rijkenberg (1971b) noted that in some hosts the mitotic activity produces layers of cells that are corklike in appearance. These investigators (Joubert & Rijkenberg 1971a) also observed that in senescing leaves, such alga-induced mitotic activity produced regions of green (i.e., nonsenescent) leaf tissue around the algal thalli. The wound response entails necrosis of subtending tissue, which, as mentioned, can be restricted to a few cell layers beneath the algal thallus or can extend from the dorsal to the ventral epidermis. In some hosts the necrosis entails the accumulation of putative suberin and tan-

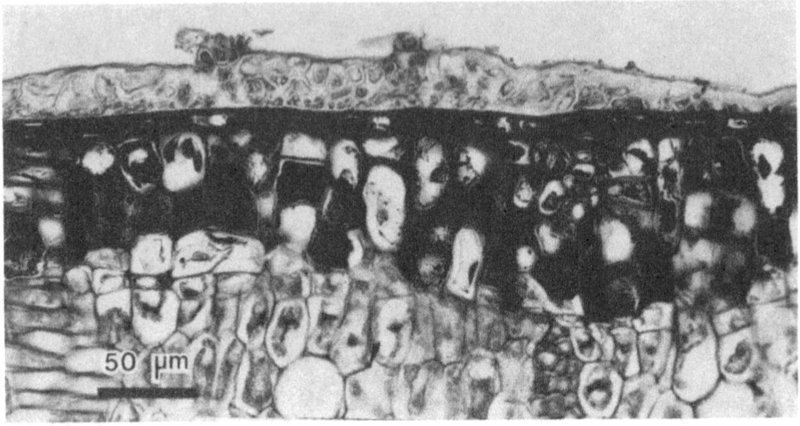

Fig. 10. Wound response in *M. grandiflora* leaf beneath *Cephaleuros* thallus. Light micrograph.

nins within the moribund cells. In those hosts exhibiting a wound response the alga may grow subcuticularly above the epidermis (Figs. 10, 11) or may penetrate into the leaf mesophyll. It is important to note that there is no clearly demonstrated example of intracellular penetration by *Cephaleuros*. In those cases where intercellular penetration occurs, it is likely that the alga simply invades the area between weakly adhering dead cells.

Without engendering extensive semantic dispute, it can be stated that *Cephaleuros* is, in some cases at least, a parasite that harms its obligate host. The amount or extent of the "harm" can be discussed and, as mentioned earlier, Thompson queried the role played by other pathogens (e.g., fungi) during severe *Cephaleuros* infections. Nevertheless, a foreign organism living under the cuticle of its vascular plant host can induce a wound reaction. The reaction may be general and the "symptoms" of a *Cephaleuros* infection (Fig. 10) might well be induced by other biological and nonbiological agents (e.g., fungi and physical tears, respectively). Some hosts exhibit wound responses to *Cephaleuros* infection that are unlike the responses (if any) to other known pathogens affecting the same host. The specific cause of *Cephaleuros*-induced wound reactions is unknown. Various authors have suggested shading, enzymes, or toxic compounds released by the alga and the loss of water, dissolved minerals, and/or organic compounds as possible direct causes of the wound response. Now, as in 1929, when F. A. Wolf reviewed the problem, there is no experimental evidence to indicate which factors are involved. Wolf did comment on the fact that shading should certainly be a deleterious influence and that *Cephaleuros* thalli remain alive "for a considerable period on fallen leaves which have been occasionally moistened." This latter

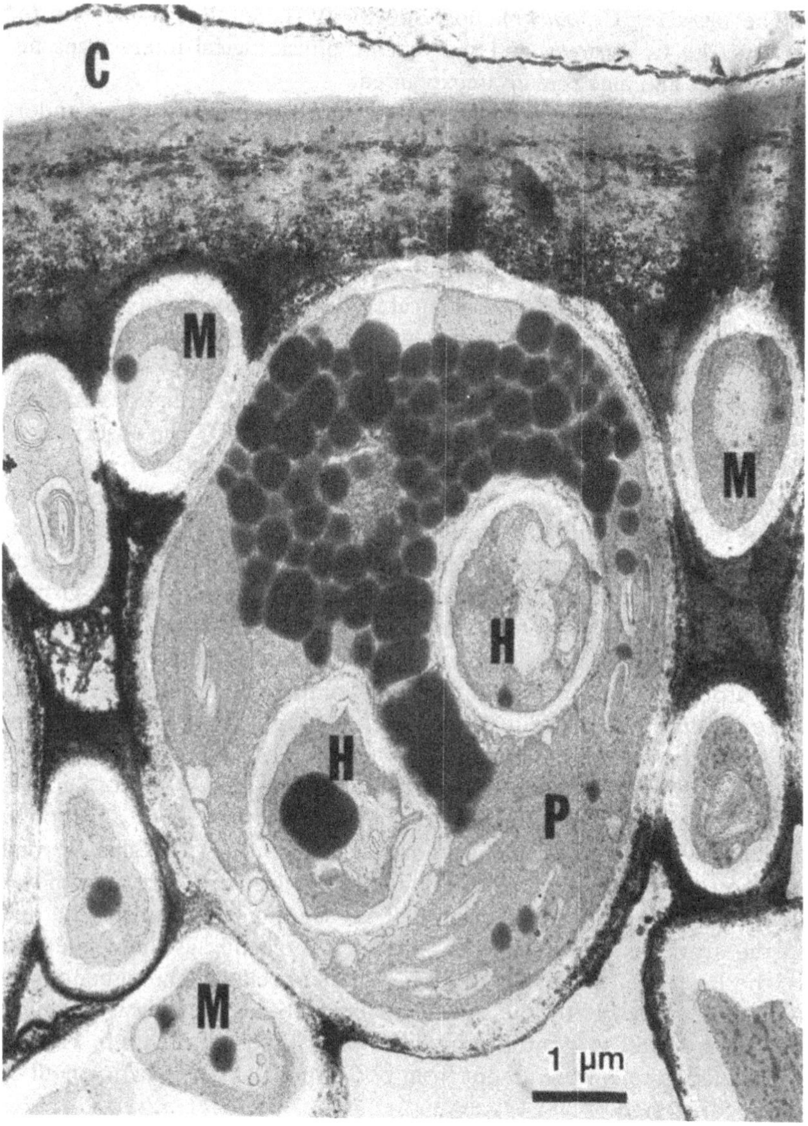

Fig. 11. *Strigula* sp. A lichenized *Cephaleuros*. C, host cuticle; P, phycobiont; H, fungal haustoria; M, mycobiont hyphae. Transmission electron micrograph.

observation and the fact that *Cephaleuros* is not highly specialized (i.e., not very host specific) led him to conclude that *Cephaleuros* was dependent on its host for water and inorganic salts only. The host, of course, also provides a habitat and, presumably, protection from desiccation for the subcuticular portion of the thallus.

The basis for *Cephaleuros* host specificity, general as it may be for species like *C. virescens*, and the specific physiological interactions between host and alga remain unexplained.

Biochemical and physiological studies are prerequisite to an understanding of the basis of host specificity in *Cephaleuros* and the exact nature of the alga-host relationship in *Cephaleuros* infections. As early as 1907, Mann and Hutchinson grew *Cephaleuros* in culture and there have indeed been several reports on this topic (e.g., Wolf 1930; Suématu 1957; Jourbert 1969; Chowdary 1969, 1970; Joubert et al. 1975; Jose & Chowdary 1979). The comparative physiological studies of Joubert and co-workers are particularly interesting (Joubert et al. 1975), and several observations should be noted. First, the addition of 1% glucose or 0.25–0.5% yeast extract to an inorganic medium (modified Chu 10) "caused a marked stimulation of growth of the algae under artificial light." The observations were extended to other inorganic media, and the growth-promoting effect of an utilizable carbon source was again observed. Second, both nitrate and ammonium ions were used by the alga as a source of nitrogen. Among those amino acids that stimulated the growth of the alga, L-proline and L-arginine produced the greatest effect. Third, heterotrophic growth of *Cephaleuros* in complete darkness was sustained by three hexoses (D-fructose, D-galactose, and D-glucose). Maltose and sucrose (α-glucosides) also promoted heterotrophic growth. Comparable stimulatory effects were produced by melezitose, a trisaccharide (glucose-fructose-glucose) found in the sap of some coniferous trees and in the exudates on the leaves of some trees (the exudate, however, may be a product of aphids or other insects rather than of the plant; Axelrod 1965). Fourth, although D-ribose and D-sorbitol did support weak growth of the alga in total darkness, there was a lag period of about 3–4 weeks before growth was detectable. The authors mention that the lag might presumably be due to the synthesis of inducible enzymes. As intriguing as the suggestion is, the lag period seems far too long for such cellular phenomenona as now understood. Fifth, there was no demonstrable hydrolysis of starch by the alga after 2 months' growth on solid Emerson medium. Finally, reproductive cells were absent from both strains of *Cephaleuros* on all of the media tested.

The preceding observation on the absence of reproductive structures in culture-grown *Cephaleuros* leads this discussion back to the work of Chowdary (1969) and Jose and Chowdary (1978) in which the formation of reproductive structures in *C. virescens* and *Trentepohlia effusa* (Krempelhuber) Hariot was induced in culture through the addition of plant hormones and growth substances. In concentrations ranging from 10^{-8} M to 10^{-4} M, IAA, IBA, and IPA induced the formation of "stalked sporangia" (i.e., zoosporangia). Strangely, indole induced the formation of both zoosporangia and gametangia (10^{-7} M and 10^{-6} M) or only gametangia (10^{-5} *M* and 10^{-4} *M*). Until confirmed or disproven experimentally, the effect of indole should be regarded as a hormone substrate effect; that is, the addition of indole feeds the biosynthetic pathway producing IAA or analogues. A direct effect by indole, which is stereochemically inactive as an auxin in all other bioassays, would be an exciting discovery. Somewhat comparable results were reported in a study of culture-grown *Trentepohlia* (Jose & Chowdary 1978) exposed to IAA, IBA, ICA, and indole at concentrations of 10^{-10} *M* to 10^{-5} *M*. Although both studies must be regarded as preliminary indications of the possible effects of exogenous hormones and growth regulators, one can scarcely resist correlating the absence of reproductive structures and the absence of plant hormones in culture-grown *Cephaleuros*.

Returning to the basic question of alga-host interaction, one can readily wish to extrapolate and hypothesize that *Cephaleuros* reproduction in nature requires host plant hormones. Some species of *Cephaleuros* do indeed seem to be obligately epiphytic in nature, and yet securing water, minerals, and subaerial habitats does not require a plant host (cf. *Trentepohlia* and *Phycopeltis*). The stimulation to both autotrophic and heterotrophic growth by utilizable carbon sources combined with the possible host hormone effects argues in favor of the hypothesis that *Cephaleuros* is more than a "water parasite" requiring only water and dissolved minerals (cf. earlier discussion of Wolf's ideas). As enthusiastic as one might be, the fact that convincing experimental evidence is not yet available precludes definitive statements.

There have been at least two attempts to demonstrate experimentally the physiological relationship between *Cephaleuros* and its hosts. Vidhyasekaran and Parambaramani studied the carbon metabolism (1971a) and the nitrogen metabolism (1971b) of alga-infected plants. In both studies, healthy and alga-infected leaves were compared. In the words of the authors, "extra care was taken to select the uniformly severely infected leaves" and "comparative healthy leaves were selected" either "randomly" in the carbon study or "scrupulously" in the nitrogen study. The major observations in the carbon study were as follows. First, in the three hosts studied (mango, guava, and sapota), the algal infection caused an increase in the total sugar content of the leaves (as determined by the anthrone reagent method of Yemm & Willis 1954). Second, only three sugars (fructose, glucose, and sucrose) were detected in both healthy and diseased leaves. Of these three, glucose and sucrose were diminished in infected tissues. Conversely, fructose levels were greatly increased in infected leaves. Third, starch content was greater in infected leaves, as was cellulose content. Fourth, although pectin content was slightly greater in infected leaves, lignin content was not significantly increased.

The nitrogen study provides an interesting comparison. First, there was a reduction in the total nitrogen content of infected leaves. Second, there were decreases in the ammoniacal and nitrite nitrogen content of infected leaves, but an increase in nitrate nitrogen. Third, although there were decreases in infected leaves of the amino, amide, and protein nitrogen content, there were increases in the levels of certain amino acids (glutamic acid, alanine, methionine, and arginine). The initiation of experimental study of the metabolism of plants infected by *Cephaleuros* is to be applauded, and experimental investigation may someday elucidate the alga-host physiological interaction. In reviewing the works cited, however, it should be noted that the intimate association between the alga and the host leaf precludes differentiation of parasite and host metabolism. That is, the carbon and nitrogen content of the alga is assayed with that of the infected leaf. The distribution of carbohydrates and nitrogenous compounds in leaf tissue before and after infection would be a more meaningful indication of the effects of the alga. Similarly, tracer studies indicating the specific movement of a variety of compounds from the host into the parasite (and vice versa) would answer many questions about the host-parasite interaction. Such experiments are feasible and will probably be undertaken sooner or later. In the absence of such studies, it is perhaps advisable to avoid extensive interpretation of the observations reported by Vidhyasekaran and Parambaramani (1971a, b). However, the fundamental observation that both the carbon and nitrogen metabolism of the alga infected leaves is altered indicates that *Cephaleuros* is an interactive symbiont.

The lichenized alga

Cephaleuros is the phycobiont in 14 species of obligately foliicolous lichens (Santesson 1952) in the genera *Strigula* and *Raciborskiella*. An ultrastructural study of *Strigula elegans* (Fée) Müller Arg. (Chapman 1976a) demonstrated fungal haustorial penetration of the phycobiont (Fig. 11). Haustoria penetrated normal vegetative algal cells and thus were not restricted to senescing or decaying cells, as occurs in other pyrenocarpous lichens (Chapman 1976a). Because the lichens occupy the same habitat as does the nonlichenized alga (in fact, the algal thallus *is* the habitat) and because some phycobiont cells are eventually destroyed by the haustorial penetration, the alga is parasitized and does not benefit from the symbiotic association. Thus, the parasitic alga is parasitized and we have a clear example of hyperparasitism. In discussing the interaction between the vascular plant host and the lichenized alga vis-à-vis the nonlichenized alga, it must be made clear that there are two distinct possibilities. First, existing *Cephaleuros* thalli can be lichenized. Second, the lichen can establish a new "infection" presumably through either heretofore underscribed lichen propagules or simultaneous, colocated algal thallus establishment, fungus growth, and immediate subsequent symbiosis. Although

Roth observed that the pathological effect of *Cephaleuros* was increased by an association with certain fungi (viz. *Alternaria* sp., *Diplodia* sp., *Botryodiplodia* sp., *Pestalotia* sp., *Helminthosporium* sp., and *Phoma* sp.), the lichenized form of the alga is generally considered innocuous. Lichenization also slows growth of the thallus.

In considering both the pathological effects of the lichen and the growth rate of the lichenized thallus, one must remember the two distinct conditions just cited. If a well-established, relatively large *Cephaleuros* thallus becomes lichenized, there may well be a large necrotic wound beneath the thallus. However, the wound would have been produced before lichenization. Similarly, the ultimate size of the lichen thallus prior to leaf abscission may be relatively large (i.e., 1 cm or more in diameter); however, most of the increase in thallus diameter may have occurred prior to lichenization. Leaves infected by lichenized *Cephaleuros* (e.g., *M. grandiflora* leaves) often bear a large number of very small thalli of relatively uniform diameter. Leaves bearing secondarily lichenized *Cephaleuros* bear fewer thalli that vary in size from barely visible to quite large.

Perhaps one of the most interesting observations heretofore reported for lichenized *Cephaleuros* is the description of host susceptibility recounted by Wellman (1972). Some host plants are susceptible to *Cephaleuros* but not to the lichen. Other hosts are susceptible to *Strigula* but not to *Cephaleuros*. Finally, some plants are susceptible to both *Cephaleuros* and *Strigula*, and others are susceptible to neither. Wellman also called attention to the genus *Congea* (an ornamental vine) in which the showy floral bracts were susceptible to *Strigula* but not *Cephaleuros*, and the true leaves were susceptible to *Cephaleuros*. Since a leaf could be subject to infection by *Cephaleuros* or *Strigula* and since these may be regarded as two distinct organisms, differences in host susceptibility are clearly understandable. However, once a leaf is infected by *Cephaleuros*, eventual formation of *Strigula* by lichenization of the algal thallus should be possible if suitable ascomycetous fungi are present (as they apparently were in some of the cases discussed by Wellman). Thus, one might expect any host susceptible to *Cephaleuros* to bear *Strigula* as well. Wellman indicates that it is both possible and probable that biologic races of *Cephaleuros* and of *Strigula* exist.

The genus *Phycopeltis*

Although *Phycopeltis* typically occurs in tropical and subtropical regions (e.g., Printz 1939, 1964; Prescott 1968; Watson 1970), it also occurs in more northerly and southerly latitudes. The type species *P. epiphyton* Millardet was reported by Millardet (1870) from northern and central Europe, and the genus has also been observed in Japan (Molisch 1926;

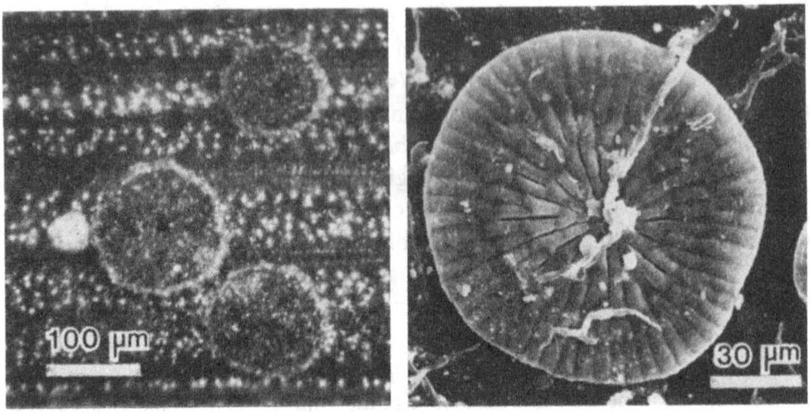

Fig. 12. Left. *Phycopeltis* thalli on palmetto palm (*Sabal minor*). Light micrograph. **Right.** *Phycopeltis* thallus supracuticular on host leaf. Scanning electron micrograph.

Suématu 1957) and New Zealand (Jennings 1896). The first published account of the genus in the continental United States was by Chapman and Good (1976).

Phycopeltis has been reported most often to be epiphytic on the leaves of higher vascular plants. *P. epiphyton* was first observed on the upper surface of gymnosperm leaves (Millardet 1870) and since that time has been reported as occurring on numerous other higher vascular plants (e.g., Good & Chapman 1978a). Although the alga typically occurs on adaxial leaf surfaces it may occur abaxially as well. In addition, *Phycopeltis* species have also been reported to epiphytize the stems and fruits of angiosperms (personal communication by R. H. Thompson), the leaves of lower vascular plants (e.g., ferns, lycopods), and bryophytes (Molisch 1926; Good, unpublished observation). In contrast to *Cephaleuros, Phycopeltis* has also been reported to occur on bark and abiotic substrates such as rocks (Scannell 1965).

Although a few *Phycopeltis* species consist of loosely arranged filaments, most species form a disc-shaped monostromatic pseudoparenchymatous thallus composed of laterally appressed dichotomously branched filaments (Fig. 12). The immature vegetative thalli are microscopic and typically grow to 1–3 mm in diameter. The coloration of thalli depends, at least in part, on environmental conditions. Growing in a shaded humid regime, the alga typically has a grass-green coloration. If the alga is exposed to higher light intensities, a deep yellow, orange-red to copper-red coloration develops due to the production and accumulation of the cytoplasmic pigment hematochrome.

Most *Phycopeltis* species form vegetative thalli that do not produce sterile hairs (trichomes), and the only heterotrichous condition ever ob-

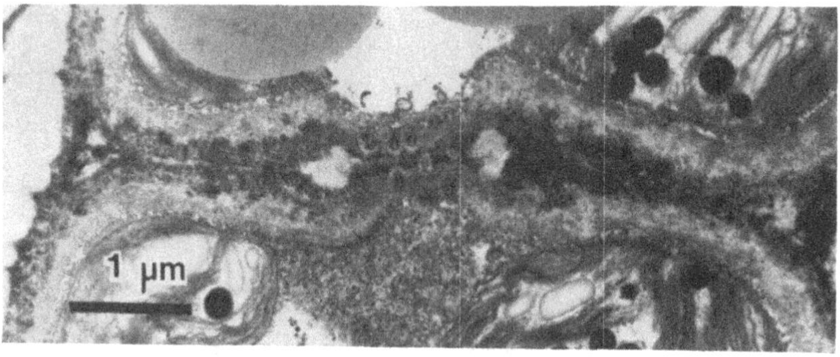

Fig. 13. Typical crosswall in vegetative cells of *Phycopeltis*. Note that the plasmodesmata are restricted to a thin central region of the crosswall. Transmission electron micrograph.

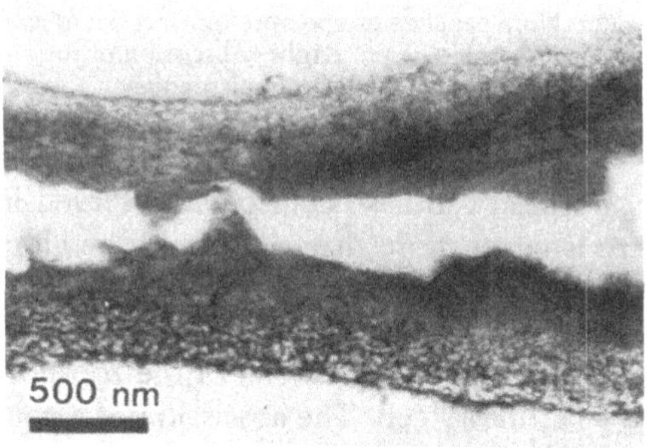

Fig. 14. Sporopollenin in acetolyzed walls of *Phycopeltis*. Transmission electron micrograph.

served is produced by the formation of highly specialized asexual reproductive structures. Rhizoids have never been reported for the genus.

Ultrastructurally, the vegetative crosswalls of *Phycopeltis* are unique compared with the other subaerial algae studied. Although simple plasmodesmata (40–50 nm diameter) localized in the central region of a septum form a central pit area (Fig. 13), a ring of light-staining material is found within the distinctly thickened border that encloses the central pit area. In addition, a sporopollenin has been reported to occur in the cell walls (Fig. 14) of at least one species of *Phycopeltis* (Good & Chapman 1978a).

Phycopeltis species reproduce asexually by the production of zoospores in highly specialized zoosporangia. In most species, any cell of the vegetative thallus has the potential to produce zoosporangia. The formation of a zoosporangium is initiated by the production of a fertile trichome from a single cell of the vegetative thallus. The distal end of the fertile branch expands and produces a single zoosporangium attached to a suf-

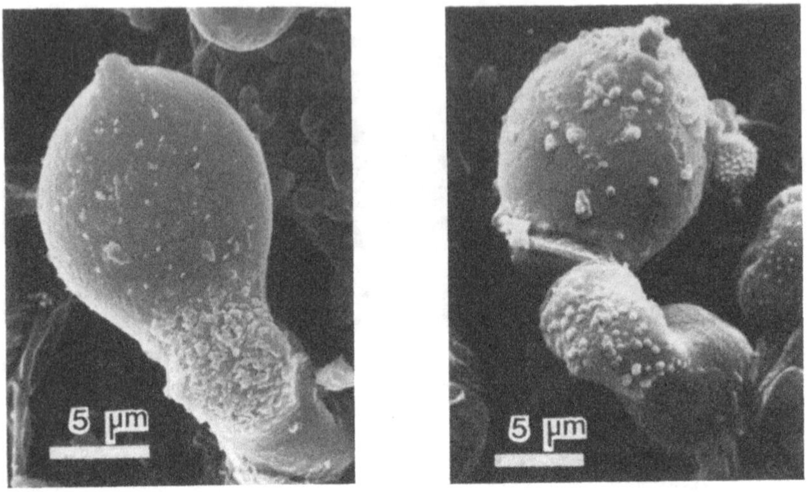

Fig. 15. Left. *Phycopeltis* zoosporangium. Note papillate escape pore (not yet open) and short stalk (cf. Fig. 5, right). Scanning electron micrograph. **Right.** Abscission of the zoosporangium in *Phycopeltis*; note escape pore. Scanning electron micrograph.

fultory cell. The production of a head cell has never been observed in this genus and as a result there is never more than one zoosporangium produced by a fertile trichome. As the zoosporangium matures, it produces a nipplelike protuberance, which is the future escape pore (Fig. 15). The escape pore is always at or near the distal end of the zoosporangium and never adjacent to a suffultory cell. The abscission of a zoosporangium from its suffultory cell is similar to the process in *Cephaleuros*.

Zoospores are produced by progressive cytoplasmic division and, as in other subaerial algae, the pectinaceous plug of material occluding an escape pore dissolves when wetted before zoospore discharge. At first there is an explosive release of zoospores followed by a slower release, during which individual zoospores can be observed to squeeze through the escape pore.

Zoospores are naked, lack an eyespot and pyrenoids, and are quadriflagellate even though they may appear biflagellate due to twisting of keeled flagella (cf. *Cephaleuros*). Although the zoospore flagellar apparatus of *Phycopeltis* (Good 1978) is similar to that observed in *Cephaleuros* (Chapman 1981), there is one distinct difference, namely, a band of electron-dense material that is positioned directly above the basal bodies. The significance of this distal band is not understood.

Sexual reproduction is the second method by which *Phycopeltis* is propagated. In most species, the gametangium is a sessile, modified vegetative cell, and any cell of the vegetative thallus can become a gametangium (Fig. 16). However, in some species only marginal cells of the thallus

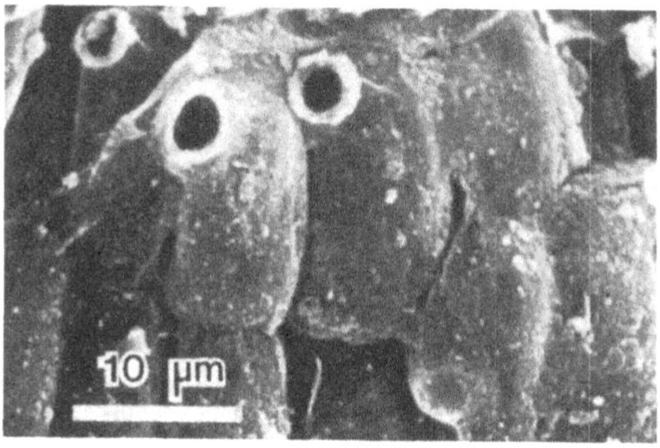

Fig. 16. *Phycopeltis* gametangia; note escape pores (cf. Fig. 7, bottom left). Scanning electron micrograph.

produce gametangia and in one species gametangia may also be produced by the cells of an erect filament. A nipplelike protuberance (i.e., the future escape pore) is produced on the upper surface of an intercalary gametangium or at the terminal end of an apical gametangium. A plug of pectinaceous material occludes the escape pore prior to gamete release.

Gametes are formed by repeated nuclear division followed by progressive cytoplasmic cleavage. Eight or more gametes may be formed (in one instance over 100 gametes were observed by Thompson). When wetted, the plug of material occluding the escape pore dissolves and an initially forceful discharge of gametes occurs. The gametes are of the same size, are biflagellate, possess keeled flagella, and lack pyrenoids or eyespots. Flagella insert into an anterior apical papilla from opposite sides and a distal band is located immediately above the two basal bodies. Fusion of gametes has not been observed, and it is not known whether the plants are monoecious or dioecious.

In the life cycle of *Phycopeltis*, zoosporangia and gametangia may occur on the same vegetative thallus. In asexual reproduction, the discharged zoospores settle down on a substrate and lose their flagella. Upon germination, the zoospores form a four-lobed immature thallus that is incompletely septate. It is not known whether there is an alternation of heteromorphic generations (including dwarf sporophyte and meiozoospores). Although there is circumstantial evidence for an alternation of isomorphic generations, the entire question of postzygotic development is still unclear and warrants further study.

As mentioned earlier, *Phycopeltis* grows on both living and nonliving substrates. This strongly implies that *Phycopeltis* is a free-living epiphyte and not a parasite. The genus has been considered strictly epiphyllous because it is easier to find on leaves and hence is probably sought and found more often on this substrate. When *Phycopeltis* does epiphytize leaves, it always grows above the leaf cuticle and does not cause any

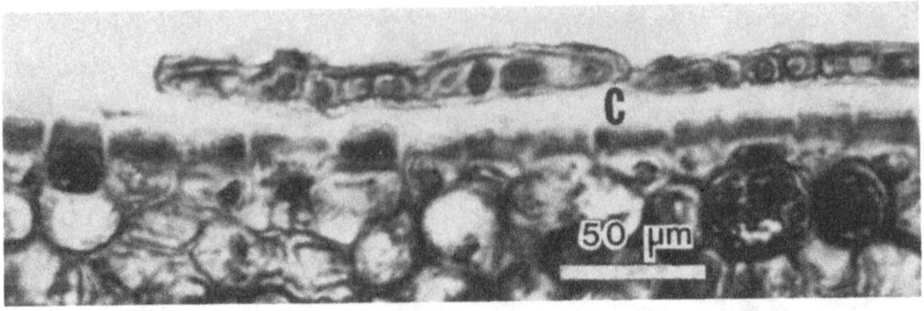

Fig. 17. *Phycopeltis* on host leaf. Note supracuticular position and the absence of a wound response in the subtending host tissue (cf. Fig. 10). C, host cuticle. Light micrograph.

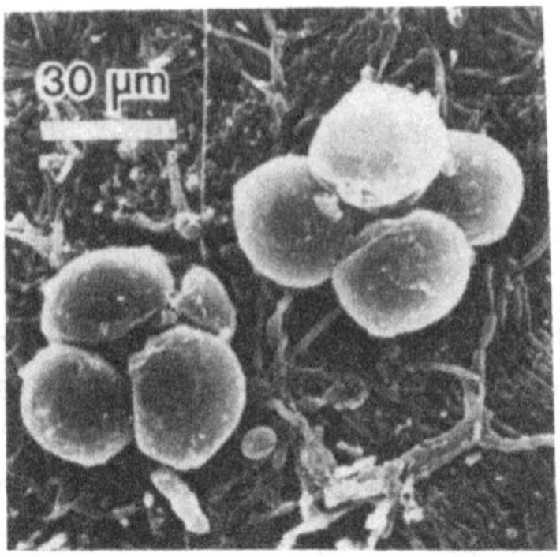

Fig. 18. *Stomatochroon* gametangia on host leaf. Scanning electron micrograph.

apparent wound reaction in subtending leaf tissues (Fig. 17). Although there are no published studies on the physiological interactions between *Phycopeltis* and its biotic hosts, the fact that this alga also grows on abiotic substrates suggests that the genus should be considered a facultative epiphyte.

The genus *Stomatochroon*

Stomatochroon is the most morphologically reduced and least known of all the Chroolepidaceae. The information available on this genus is at best scanty, and published ultrastructural and biochemical studies are lacking. The genus was first described by Palm (1934) on one species, *S. lagerheimii*, and was reported to occur on many different angiosperms in tropical and subtropical regions. Although seen in Florida by Thompson, the first published report of *Stomatochroon* from the continental United States was from Louisiana (Good & Timpano 1980).

Stomatochroon species inhabit the stomata of leaves (Fig. 18) and are thus "stomaticolous." Although the alga may inhabit adaxial stomata, it

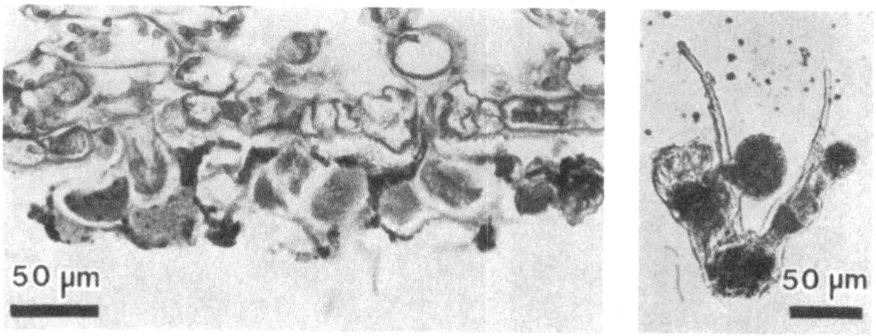

Fig. 19. Left. *Stomatochroon* thalli in host abaxial stomata. Light micrograph. **Right.** Isolated, intact *Stomatochroon* thallus. Note zoosporangium and sterile trichomes. Light micrograph.

characteristically is found in stomata on the ventral surface (Fig. 19, left side). The genus is most readily seen on the drip tips of leaves. The portion of the alga above the leaf surface may vary in color from green, orange, and red to purple due to environmental conditions (e.g., light intensity and humidity).

The vegetative thallus (Fig. 19, right side) consists of a rounded or ellipsoidal central thallus cell (basal cell) that is in a stoma and protrudes from the stomatal aperture. The central thallus cell produces one or more long thin sterile trichomes and within the substomatal chamber, a lobed anchor cell that helps secure the alga. In other species (as yet unpublished but observed by the authors and described in detail by Thompson in a personal communication), there is an intercellular system of branching filaments in the mesophyll of the host leaf. In some cases these filaments may be poorly developed, possess cells that have thin walls, and appear colorless to pale green due to a reduced number of chloroplasts. In others, they may be well developed, possess cells with thick walls, and appear green due to normal chloroplasts. The tip of an intercellular filament may enter a stoma and initiate new algal growth above the leaf surface.

Stomatochroon reproduces asexually by the formation of zoospores in highly specialized zoosporangia. During zoosporangium formation the central thallus cell produces one or more reproductive branches (i.e., fertile trichomes). Typically, each reproductive branch produces a head cell (cf. *Cephaleuros*) and each head cell produces one or more suffultory cells, each of which bears a zoosporangium (Fig. 20, left side). The zoosporangium produces a single escape pore that is plugged with pectinaceous material. The abscission of zoosporangia from suffultory cells is the same as in *Cephaleuros* and *Phycopeltis*. The plug of pectinaceous material dissolves when exposed to moisture, and zoospores (typically 8–16) are discharged through the escape pore. Zoospores lack eyespots

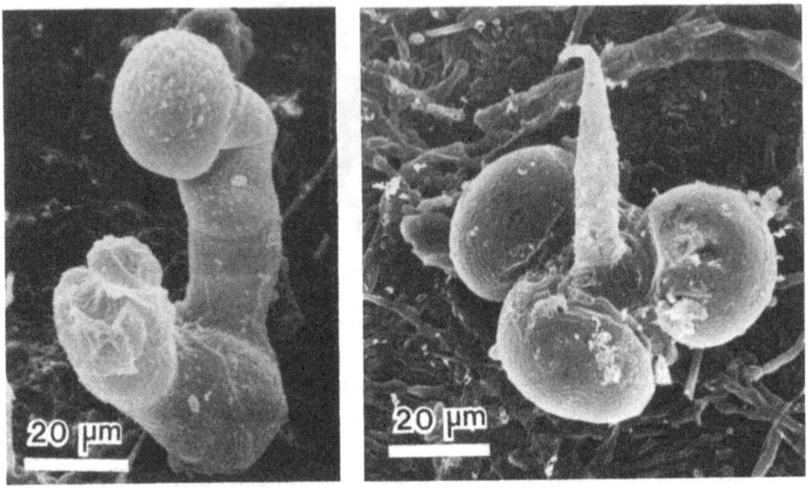

Fig. 20. Left. *Stomatochroon* zoosporangium on sporangiophore. Scanning electron micrograph. **Right.** Gametangia and sterile trichome of *Stomatochroon*; stoma is obscured by the alga. Scanning electron micrograph.

and are quadriflagellate although they may appear to be biflagellate (cf. the zoospores of *Cephaleuros* and *Phycopeltis*).

Stomatochroon also reproduces sexually. Rounded or ellipsoidal gametangia (usually two to eight) are produced by the central thallus cell and the gametangia are located above the stoma (Fig. 20, right side). Each gametangium produces the typical plugged escape pore, which dissolves when wetted, and gametes are usually discharged explosively. Gametes are biflagellate and do not have an eyespot. Within 2–5 min after release, clusters of many gametes can be seen on the alga or near it. Presumably syngamy occurs within clusters of gametes within a gametangium.

The life cycle of *Stomatochroon* has been reported by Thompson (1959). During asexual reproduction released zoospores come to rest and germinate, growing into a stoma (Thompson, personal communication). Sexual reproduction is similar to that in *Cephaleuros*. *Stomatochroon* is isogamous, homothallic, and possesses an alternation of heteromorphic generations. The reduced sporophytic generation typically consists of a two-celled sporophyte (a dwarf sporophyte). In the production of a dwarf sporophyte, a zygote settles on the alga or leaf surface and upon germination develops into a reproductive branch. Although details may vary, the reproductive branch ultimately produces a suffultory cell to which is attached a sporangium. This sporangium is considered a meiozoosporangium although there is no direct evidence (e.g., cytological, cytochemical) that meiosis occurs here. Meiozoospores escape through an escape pore and are quadriflagellate. Presumably meiozoospores germinate in a stoma.

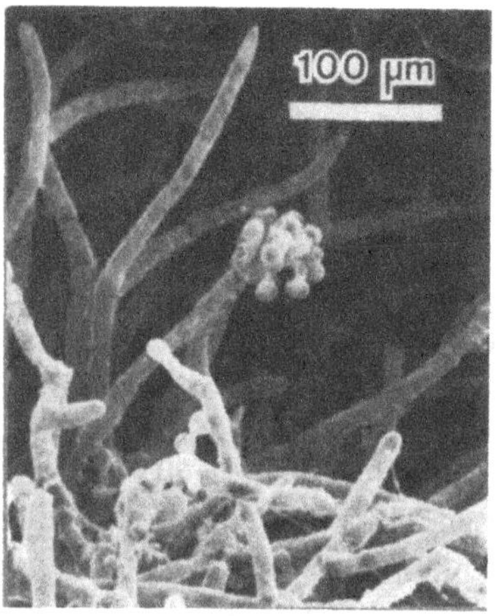

Fig. 21. Sporangiophore and zoosporangia of *Trentepohlia* (cf. Fig. 5, right). Scanning electron micrograph.

From the small amount of information available, it appears that *Stomatochroon* is restricted to the stoma of leaves and the intercellular spaces between mesophyll cells. The genus has never been reported to infect the fruits or stems of angiosperms and also has never been reported to occur on abiotic substrates. A distinct wounding response apparently does not occur in leaf mesophyll cells, although the tips of leaves that are heavily infected with *Stomatochroon* may turn brown. Presumably this browning of leaf host is due to the occluding of stomata. The penetration of intercellular filaments of *Stomatochroon* into host mesophyll cells has not been reported. It may be of interest to note that although the central thallus cell and reproductive structures of *Stomatochroon* are frequently absent during winter months, intercellular filaments are present in the mesophyll of the leaf. It seems likely that these intercellular filaments represent an overwintering stage that will start new growth on the leaf surface when environmental conditions become favorable. With the information available at the present time, *Stomatochroon* should be considered an obligate endophyte (parasite) of the leaves of higher vascular plants. The precise physiological relationships between the alga and its hosts must await further investigation.

The genus *Trentepohlia*

There is some irony in the fact that although *Trentepohlia* is probably the best known of the four genera of Chroolepidaceae, it will be afforded little coverage in this chapter. *Trentepohlia* forms dense tufts of branched filaments and shares many features with the other genera, including the unique zoosporangia (Fig. 21) and, of course, a subaerial habitat. *Trentepohlia* can grow on biotic and abiotic substrates and is thus a facultative epiphyte on its vascular plant hosts. Unlike *Phycopeltis, Trentepohlia* is generally quite visible wherever it occurs and is frequently collected from rocks and wooden boards. *Trentepohlia* growing on the bark of trees does not induce a wound response and does not invade the host tissue. There are no reports of wound responses in leaves that bear *Trentepohlia*, nor are there reports of the girdling of young stems by *Trentepohlia* infections. The only interaction between alga and host warranting attention is the damage caused by luxuriant growth of *Trentepohlia* on leaves. In such instances the leaves are heavily shaded and suffer accordingly. Even when responsible for deleterious effects on the host leaves, the epiphyte does not penetrate the injured tissue and should not be termed a parasite. It is of interest that both *Trentepohlia* and *Phycopeltis* do not require vascular plant hosts and both are more widely distributed than the other two genera, ranging into temperate climatic regions (cf. the tropical and subtropical distribution of *Cephaleuros* and *Stomatochroon*).

The Chroolepidaceae: a continuum of interaction strategies

The four genera discussed in this chapter demonstrate a continuum of interaction strategies involving increasing dependence on vascular plant hosts. *Trentepohlia* grows well on either abiotic or biotic substrates and seems to interact very little, if at all, with its plant hosts. *Phycopeltis* is similarly facultative in its growth on plant hosts, but it is certainly found most commonly on plant hosts rather than on abiotic substrates. The genus also exhibits greater morphological differentiation than does *Trentepohlia*. One might argue that the discoidal thallus is an adaptation to epiphytic growth on more or less smooth substrates. A comparison can be made to *Coleochaete scutata* Brébisson, which is also epiphytic and similar in gross morphology.

Cephaleuros retains a discoidal thallus, but also includes features that can be cited as special adaptations for its subaerial, parasitic existence. It exhibits advanced heterotrichy (including rhizoidal filaments, sporangiophores, and abundant sterile hairs). Second, *Cephaleuros* is an obligate symbiont. Third, there may even be a physiological dependence on the host for exogenous hormones or growth substances. Located further along the continuum of interaction strategies, *Cephaleuros* also induces pronounced, deleterious effects in some hosts. Finally, *Stomatochroon* can be described as a highly reduced form that has moved from the primarily supraepidermal habitat occupied by *Cephaleuros* to an intercellular home within the mesophyll.

Is the continuum of interaction strategies also an evolutionary continuum? It could be, and there is thus reason to seek the ultrastructural and biochemical data that will elucidate the phylogeny of these four genera.

Some notes on other green algae

In their comprehensive review of parasitic green algae, Joubert and Rijkenberg (1971b) provided detailed coverage of the genus *Cephaleuros* and included pertinent information on several other genera. The following notes are proferred as a short addendum, both to this chapter and to their review. *Chlorochytrium* is an endophyte of hydrophytes (e.g., *Lemna* spp., *Hypnum* spp., and *Sphagnum* spp.) and in some cases (e.g., *C. inclusum*; Kornmann 1972) may be an endophytic stage of a free-living alga (Kornmann 1972; Ralph A. Lewin, personal communication). Thompson (1970) stated that *Chlorochytrium* has not only an endophytic existence, but also a neustonic and an epiphytic existence as well. Therefore, *Chlorochytrium* is not an obligate endophyte. The coenocytic genus *Phyllobium* was recently collected on *Eichhornia crassipes* (Srivastava & Noor 1978); however, the alga has not been extensively collected or studied, and accordingly little information concerning its interaction with the host plants is available. *Rhodochytrium* grows in angiosperms, such as *Ambrosia* and *Asclepias*, and, because it is nonphotosynthetic, can be considered "advanced" in the evolution of parasitic green algae. The effects of the alga on its host plants range from little overall damage to definite stunting. There apparently has been little published research on this unusual and controversial alga. *Phyllosiphon* is a branched coenocyte that occupies the intercellular spaces of leaves of host plants in the Araceae. Although there has apparently been some question about both the presence of chlorophyll in the vegetative filaments and the divisional placement of the taxon, spectrophotometric analysis (Leclerc & Coute 1976) has demonstrated the presence of both chlorophylls *a* and *b*. Despite emphasis on the value of a study of the physiology and ultrastructure of *Phyllosiphon* by Joubert and Rijkenberg (1971b), few published data have appeared. Dr. K. A. Pirozynski (National Museums, Ottawa) has sought a source of fresh material for the study of this alga (personal communication) and investigation may be underway by him. The last genus mentioned by Joubert and Rijkenberg (1971b) was the monotypic *Phytophysa* Weber-van Bosse, an organism the authors characterized as an alga "on which no further work has apparently been done since its first description in 1890." Their characterization remains appropriate.

References

Axelrod, B. (1965). Mono- and oligosaccharides. *In: Plant Biochemistry*, ed. J. Bonner & J. E. Varner, p. 250. New York: Academic Press.

Batista, A. C., & Lima, D. A. (1949). Lista de suscetiveis da alga *Cephaleuros mycoidea* Karst em Pernambuco. *Boletim Sec. Agr. Ind. Com.* **16**, 32–46.

Blinn, D. W., & Morrison, E. (1974). Intercellular cytoplasmic connections in *Ctenocladus circinnatus* Borzi (Chlorophyceae) with possible ecological significance. *Phycologia* **13**, 95–7.

Bourrelly, P. (1966). *Les Algues d'Eau Douce. Tome I. Les Algues Vertes*. Paris: N. Boubee et Cie.
Chapman, R. L. (1976a). Ultrastructural investigation on the foliicolous pyrenocarpous lichen *Strigula elegans* (Fée) Müll. Arg. *Phycologia* **15**, 191–6.
– (1976b). Ultrastructure of *Cephaleuros virescens* (Chroolepidaceae; Chlorophyta). I. Scanning electron microscopy of zoosporangia. *Amer. J. Bot.* **63**, 1060–70.
– (1980). Ultrastructure of *Cephaleuros virescens* (Chroolepidaceae; Chlorophyta). II. Gametes. *Amer. J. Bot.* **67**, 10–17.
– (1981). Ultrastructure of *Cephaleuros virescens* (Chroolepidaceae; Chlorophyta). III. Zoospores. *Amer. J. Bot.* **68**, 544–56.
Chapman, R. L., & Good, B. H. (1976). Observations on the morphology and taxonomy of *Phycopeltis hawaiiensis* King (Chroolepidaceae). *Pac. Sci.* **30**, 187–95.
– (1978). Ultrastructure of plasmodesmata and cross walls in *Cephaleuros, Phycopeltis* and *Trentepohlia* (Chroolepidaceae; Chlorophyta). *Br. Phycol. J.* **13**, 241–6.
Chappell, D. F., Stewart, K. D., & Mattox, K. R. (1978). On pits and plasmodesmata of trentepohlialean algae (Chlorophyta). *Trans. Amer. Micros. Soc.* **97**, 88–94.
Chowdary, Y. B. K. (1969). Induction of reproductive organs in *Cephaleuros virescens. Indian J. Microbiol.* **8**, 153–8.
– (1970). Cultural and nutritional requirements of *Cephaleuros virescens* Kunze. *Indian Biol.* **2**, 75–9.
Chowdary, Y. B. K., & Jose, G. (1979). Biology of *Cephaleuros* Kunze in nature. *Phykos* **18**, 1–9.
Feige, G. B., & Kremer, B. P. (1980). Unusual carbohydrate pattern in *Trentepohlia* species. *Phytochemistry* **19**, 1844–5.
Flint, E. A. (1959). The occurrence of zoospores in *Physolinum* Printz. *New Phytol.* **58**, 267–70.
Fritsch, F. E. (1965). *The Structure and Reproduction of the Algae*, vol. 1. Cambridge: Cambridge University Press. (First published in 1945.)
Golato, C. (1970). Una grave malattia dell'annacardio (*Anacardium occidentale* L.) in Tanzania. *Riv. Agric. Subtrop.* **64**, 334–40.
Good, B. H. (1978). "Ultrastructural and biochemical studies on the epiphytic subaerial green alga *Phycopeltis epiphyton* Millardet." Ph.D. dissertation, Louisiana State University.
Good, B. H., & Chapman, R. L. (1978a). The ultrastructure of *Phycopeltis* (Chroolepidaceae; Chlorophyta). I. Sporopollenin in the cell walls. *Amer. J. Bot.* **65**, 27–33.
– (1978b). Scanning electron microscope observations on zoosporangial abscission in *Phycopeltis epiphyton* (Chlorophyta). *J. Phycol.* **14**, 374–6.
Good, B., & Timpano, P. (1980). Preliminary observations on the subaerial green alga *Stomatochroon. J. Phycol.* **16** (suppl.), 14.
Graham, L. E., & McBride, G. M. (1975). The ultrastructure of multilayered structures associated with flagellar bases in motile cells of *Trentepohlia aurea. J. Phycol.* **11**, 86–96.
Hansgirg, A. (1886). Prodromus der Algenflora von Bohmen, pt. 1, no. 1. *Arch. naturw. Landes Forsch. Bohm.* **5(6)**.
Holcomb, G. E. (1975). Hosts of the alga *Cephaleuros virescens* in Louisiana. *Ann. Proc. Am. Phytopath. Soc.* **2**, 134.
Jennings, A. V. (1896). On two new species of *Phycopeltis* from New Zealand. *Proc. Royal Irish Acad.* **3(3)**, 753–66.

Jose, G., & Chowdary, Y. B. K. (1978). Effect of some growth substances on *Trentepohlia effusa* (Kremp.) Hariot. *Indian J. Plant Physiol.* **21**, 6–33.
– (1979). Effect of three nitrogen sources on the growth of *Cephaleuros* Kunze isolates. *Phykos* **18**, 69–72.
Joubert, J. J. (1969). Cultivation of *Cephaleuros virescens* Kunze on an artificial medium. *Revista Biol.* **7**, 1–6.
Joubert, J. J., & Rijkenberg, F. H. J. (1971a). Studies on the host range of *Cephaleuros* spp. in Natal. *Revista Biol.* **7**, 185–93.
– (1971b). Parasitic green algae. *Ann. Rev. Phytopath.* **9**, 45–64.
Joubert, J. J., Rijkenberg, F. H. J., & Steyn, P. L. (1975). Studies on the physiology of a parasitic green alga *Cephaleuros* sp. *Phytopath. Z.* **84**, 147–52.
Karsten, G. (1891). Untersuchungen über die Familie der Chroolepideen. *Ann. Jard. Bot. Buitenz.* **10**, 1–66.
Kornmann, P. (1972). Les sporophytes vivant en endophyte de quelques Acrosiphoniacées et leurs rapports biologiques et taxonomiques. *Soc. Bot. Fr. Mem.* 75–86.
Leclerc, J. C., & Coute, A. (1976). Revision de la position systematique de l'algue parasite *Phyllosiphon arisari* Kuhn d'apres l'analyse spectrophotometrique de ses chlorophylles. *C. R. Acad. Sc. D Sc. Nat.* **282**, 2067–70.
Mann, H. H., & Hutchinson, C. M. (1907). *Cephaleuros virescens* Kunze, the "red rust" of tea. *Mem. Dept. Agr. India Bot. Ser.* **1**, 1–35, plates 1–8.
Marche-Marchad, J. (1976). Quelques aspects de l'ultrastructure de la cellule végétative et des zoïdes de deux especes de Trentepohliacées du genere *Cephaleuros* (Chlorophycées). *Bull. I. F. A. N.* **38**, 469–86.
– (1977). Observations sur la morphogenese du *Cephaleuros virescens* Kunze Chlorophycées, Trentepohliale. *Bull. I. F. A. N.* **39**, 1–22.
Marlatt, R. B., & Campbell, C. W. (1980). Susceptibility of *Psidium guajava* selections to injury by *Cephaleuros* sp. *Plant Dis.* **64**, 1010–11.
Millardet, M. A. (1870). De la germination des zygospores dans les genres *Closterium* et *Staurastrum* et sur un genre noveau d'algues chlorosporees. *Mem. Soc. Sc. Nat. Strasbourg* **6**, 37–50.
Molisch, H. (1926). *Mycoidea parasitica* Cunningham, eine parasitische und *Phycopeltis epiphyton* Millard., eine epiphylle Alge in Japan. *Sc. Rep. Tohoku Imp. Univ., Sendai, Japan* **4**, 111–17.
Palm, B. T. (1934). On parasitic and epiphyllous algae. *Arkiv Bot.* **25**, 1–16.
Papenfuss, G. F. (1962). On the circumscription of the green algal genera *Ulvella* and *Pilinia*. *Phykos* **1**, 6–12.
Prescott, G. W. (1968). *The Algae: A Review*, 2nd ed. Boston: Houghton Mifflin.
Printz, H. (1939). Vorarbeiten zu einer Monographie der Trentepohliaceen. *Nytt Mag. Naturv.* **80**, 137–210.
– (1964). Die Chaetophoralen der Binnengewasser. Eine systematische Übersicht. *Hydrobiologia* **24**, 1–76.
Rabenhorst, L. (1868). Flora europaea algarum aquae dulcis et submarinae, Sectio III. Algas chlorophyllophyceas, melanophyceas et rhodophyceas complectens. Leipzig.
Rijkenberg, F. H. J., Joubert, J. J., & Milford, K. T. (1971). The ultrastructure of the vegetative thallus of a plant-parasitic alga, *Cephaleuros virescens* Kunze. *Proc. So. African Elec. Micr. Soc.*, 27–8.
Roth, G. (1971). An algal leafspot disease on avocado pears (*Persea americana* Mill.) in South Africa. *Phytopath. Z.* **70**, 323–34.
Santesson, R. (1952). Foliicolous lichens. I. A revision of the taxonomy of the obligately foliicolous, lichenized fungi. *Symb. Bot. Upsala* **12**, 1–590.

Scannell, M. J. P. (1965). *Phycopeltis*, a genus of alga not previously recorded from the British Isles. *Irish Naturalists' J.* **15**, 75.
Smith, G. M. (1950). *The Fresh-water Algae of the United States*, 2nd ed. New York: McGraw-Hill.
Srivastava, M., & Noor, M. N. (1978). A new record of *Phyllobium sphagnicolum* Klebs from India. *Current Sc.* **47**, 123–4.
Suématu, S. (1957). Noted on *Cephaleuros* and *Phycopeltis*, parasitic and epiphytic aerial-algae. III. Lists of infected plants. *Bot. Mag. Tokyo* **70**, 276–81.
Swingle, W. T. (1894). *Cephaleuros mycoidea* and *Phyllosiphon*, two species of parasitic algae new to North America. *Proc. Amer. Assoc. Adv. Sc.* **42**, 260.
Thompson, R. H. (1959). The life cycles of *Cephaleuros* and *Stomatochroon*. *Proc. 9th Int. Bot. Congr. (Montreal)* **2**, 397.
– (1970). The genera *Chlorochytrium* Cohn and *Scotinosphaera* Klebs. *J. Phycol.* **6** (suppl.), 3.
Vidhyasekaran, P., & Parambaramani, C. (1971a). Carbon metabolism of alga infected plants. *Indian Phytopath.* **24**, 369–74.
– (1971b). Nitrogen metabolism of alga infected plants. *Indian Phytopath.* **24**, 500–4.
Watson, R. (1970). Distribution of epiphytic algae on palm fronds. *In*: *A Tropical Rain Forest: A Study of Irradiation and Ecology at El Verde, Puerto Rico*, book 2, sect. D-E, ed. H. T. Odum, pp. 233–6. Oak Ridge, Tenn.: United States Atomic Energy Commission.
Wellman, F. L. (1965). Pathogenicity of *Cephaleuros virescens* in the neotropics. *Phytopathology* **55**, 1082.
– (1972). *Tropical American Plant Disease (Neotropical Phytopathology Problems)*, chap. 26, pp. 639–68. Metuchen, N.J.: Scarecrow Press.
Went, F. A. C. (1895). *Cephaleuros coffeae*, eine neue parasitische Chroolepidee. *Centralblatt Bacteriologie* **1**, 681–7.
Winston, J. R. (1938). Algal fruit spot of orange. *Phytopathology* **28**, 283–6.
Wolf, F. A. (1930). A parasitic alga, *Cephaleuros virescens* Kunze, on citrus and certain other plants. *J. Elisha Mitchell Sc. Soc.* **45**, 187–205.
Yemm, E. W., & Willis, A. J. (1954). The estimation of carbohydrates in plant extracts by anthrone. *Biochem. J.* **57**, 508–14.

Taxonomic index

Author index

Subject index